- The 1990s will see two whole Europes, East and West, added in just ten years. Two United Kingdoms every fourteen months. A Sweden or two New Zealands per month. A Birmingham every four days. A school class of thirty every ten seconds.

- The average person in a developed country emits roughly twenty times more water and climate pollutants than their counterpart in the South. If our consumption and waste output levels do not change, the 57·5 million extra Northerners expected during the 1990s will pollute the globe more than the extra 911 million Southerners.

- In the 1980s, city populations grew by 63 million a year and outstripped all attempts to provide jobs, housing, water and sanitation. In the first quarter of the next century, cities will be adding an extra 92 million people a year.

- The ocean is our garden pond: the annual fish catch already exceeds the sustainable yield of the world's seas.

- Food production fell behind population growth in 69 out of 102 developing countries between 1978 and 1989.

- Some thirty-nine developing countries are already using water at a rate that causes regional problems. Many of them will be forced to choose between domestic, industrial and agricultural uses.

- In the first five decades of the twentieth century, one bird or mammal species was becoming extinct every 1·1 years – 230 times faster than the pre-human rate. One mammal species in every fifty has become extinct in the past 400 years.

- Every year an area of tropical forest the size of two-thirds of the United Kingdom is cleared. Every ten weeks a Netherlands. Every day a Barbados.

- Without soil conservation, the developing world could lose 18 per cent of its potential rainfed cropland and 29 per cent of potential food production.

THE THIRD REVOLUTION

THE THIRD REVOLUTION

Environment, Population and
a Sustainable World

PAUL HARRISON

I.B. Tauris & Co Ltd
Publishers
London   New York

In association with THE WORLD WIDE FUND FOR NATURE

Published in 1992 by
I.B. Tauris & Co Ltd
110 Gloucester Avenue
London NW1 8JA

175 Fifth Avenue
New York
NY10010

In the United States of America
and Canada distributed by
St Martin's Press
175 Fifth Avenue
New York
NY 10010

Published by arrangement with Penguin Books Ltd

A full CIP record for this book is available from the British Library

Library of Congress Catalog car number: 91–68020
A full CIP record is available from the Library of Congress

ISBN 1–85043–501–4   20025478

Cover photo. The dilemma at the grassroots:
Zafindraibe, desperate for land to feed his
growing family, burns down part Ranomafana
forest, Madagascar, home of the last 400
golden bamboo lemurs on earth. Photo © Paul Harrison.

304.2

For Alex and Sam
with love,
and hope

Claudius is swift in the commission of evil:
there is no chink of delay between thought and deed.
But Hamlet:
Hamlet knows from the outset that something is wrong.
By the end of act one, he knows exactly what is wrong.
By the end of act two, he knows what needs doing.
Act three brings his best chance of killing Claudius with least
    damage.
He lets it pass.
Then Polonius, Rosencrantz, Guildenstern, Ophelia, Laertes, and
    Gertrude
all die unnecessarily.
Hamlet waits until circumstances force his hand.
Before he does what had to be done all along,
Hamlet has less than half an hour to live.

CONTENTS

vii

PREFACE

As usual I have bitten off a chunk of human experience that is far too large for anyone to digest fully. My excuse is that all the areas I have covered hang together, and we must start trying to learn to look at them all together.

I have separated out my five village case studies, rather than slotting them in bits in the relevant sections, for analytical as well as aesthetic reasons. A community is not a conglomeration of unrelated elements – everything works together. The sequence of rural cases follows the progression from hunter-gatherers doing a little shifting cultivation (in Malaysia), to permanent farmers growing several crops a year (in Bangladesh).

No one is free of some sort of ideological baggage, and I don't claim to be. But the largest suitcase in my luggage is marked Empiricism: to see the world as it is, in all its complexity, to avoid preconceived ideas as far as possible, to look at the evidence, and see where it leads.

If I attack some of the arguments of those who support greater equality in the world, it's not because I disagree with the goal, it is merely because I believe that these particular arguments are faulty, in conflict with the evidence, and misguided. In some cases radicals may find themselves defeating their own objects. To say that population growth is 'not the problem' implies that it is not a problem. It provides ammunition for all those who wish to deny women the right to determine their own fertility. It also ignores the way in which population growth aggravates all those problems that are accepted as 'real' problems.

But there are equal dangers in overstating the case and saying that population growth is *the* problem. This carries the implication that other problems are less pressing. Population growth never works in isolation but always in combination with consumption levels and technology, and all the other factors that influence these three. The obsession with population above all other problems has been used (though, thankfully, very rarely) to justify the use of coercion in family planning. Coercion is just as wrong as denial of the right of access to family planning. And if we focus too narrowly on numbers alone, if we neglect all those other factors like health, education and women's status, which synergize with family planning, we shall make slower progress.

It is always tempting to oversimplify, to polarize, to reduce everything to one or two sets of polar opposites, light versus dark, good versus bad. Simplification reduces the work required of the brain. It satisfies our desire to belong to a club, to carry a banner or a label. It provides good headlines, makes good speeches.

But the world is complex. We have to simplify to some extent. But if we oversimplify, there is no hope of understanding the world properly, no hope of co-operating in tackling its many problems. If we don't co-operate, if we tackle only a few items and leave others to run rife, we won't get very far.

The chapter headings, where possible, are from my favourite piece of literature, *Hamlet*. Hopefully we will not push things quite so close to the brink as he did.

ACKNOWLEDGEMENTS

Many people have helped in making this book possible.

The main funding for the research came from the Swedish International Development Agency and the International Union for the Conservation of Nature. Travel costs were met by the United Nations Population Fund. Additional grants came from the World Wide Fund for Nature and the International Planned Parenthood Federation. I am particularly grateful to Dr Mats Segnestam, Dr Nellie van den Oever, Alex Marshall, Ivan Hattingh and John Rowley. There were no editorial strings attached, so what follows does not represent the policies of any of these organizations.

A number of other people were particularly helpful in giving information or reviewing chapters. These include: Nikos Alexandratos, John Bongaarts, Jelle Bruinsma, Mark Collins, Barry Commoner, Alan Grainger, Larry Heligman, John Jackson, Gayl Ness, Christine Ogden, Jack Parsons, David Satterthwaite, Michael Stocking, Anthony Webster and James Ypsilantis. In addition, I have a clear intellectual debt to Ester Boserup's theories, which are here modified and extended.

The Jabatan Orang Alsi in Malaysia, Patricia White in Madagascar, Rafiq Alam in Bangladesh, Tevia Abrams in India and Oxfam's Mike Butcher in Burkina Faso were particularly helpful with travel and local arrangements.

And as always I am particularly grateful to my village informants in Musoh, Ambodiaviavy, Kalsaka, Abidjan, and Hatia Island, especially Lina and Basuloh, Zafindraibe, Jean-Marie and Ninsabla Sawadogo, and Rafiq Alam. May each of them – and each of us – win their respective battles to preserve, or regain, balance with their environments.

PROLOGUE IN THE FOREST:
Musoh, Malaysia

Musoh lies deep in the central highlands of Malaysia. It is four hours' walk from the nearest road, up a steep-sided valley with slopes carpeted in forest. Wisps of hill fog drift through the tallest trees, silhouetted along the ridges, their canopies caped in trailing climbers.[1]

There is no sign of human presence in the valley, save for a thin pall of wood smoke, bluer than the rest, far up towards the watershed.

As you get closer, the forest thins. There are open patches of shrubs where butterflies flitter and feed. Then a shaky bridge of rope and slats over a rocky torrent. The village itself is no more than a score of thatch huts on wooden stilts, widely scattered among shade trees.

I stayed in the house of Basuloh, an old Semai tribesman who wore garlands of coloured paper in his hair. The interior was sparse: floor of split bamboo, walls of woven rattan, no furniture save shelves for pots and pans. One big room served as communal lounge and bedroom. At night sleeping mats were spread out and fourteen people slept side by side: Basuloh and his wife, a teenage son, plus two married sons and their wives and children. Two unmarried daughters had their own separate rooms.

The Semai are at the earliest stage of agriculture. They clear and burn small plots of forest, where they grow hill rice interspersed with cassava. After a year they plant shrub crops. Then they allow the forest to take over for ten or twenty years before they farm the same plot again. They also keep chickens and goats. But hunting and gathering still provide most of their leafy vegetables and animal protein.

On the second day of my stay, eight of the village women

gathered with rattan baskets on their backs, and set off over the
river, up hill and down dale for a mile. Two dogs came along in
case we met leopard or sunbear – there are even a few tigers left
in the hills. Every now and then they stopped to garner wild
leaves of various kinds. Finally they arrived at their current
garden. The previous year this had been cleared to grow rice.
Next year it would revert to forest for ten or twenty years. For
the interim it looked like an abandoned allotment, waist high in
weeds and twisting climbers. The women scraped the soil and
wrenched out tubers and roots, gathered chillies, cut lengths of
sugar cane. In three hours they harvested enough food to last a
week.

Hunting and fishing were men's work. The next day
Basuloh's eldest son, Baboi, took me on a hunting trip. The
forest was criss-crossed with narrow footpaths. Where the
terrain was too rough, the river itself became the path.

There were dozens of traps to check: noose traps baited with
banana to catch palm civet; square traps of twigs over the exit
holes of moonrat burrows; fierce traps of taut sharpened
saplings, ready to spear the flanks of jungle pigs.

At one deep pool, two of Baboi's sons were diving for fish
with home-made spear guns made out of wood and old tyres,
with bike spokes for darts.

Baboi had his blowpipe. From time to time he would stop
and motion me to be silent. Then he would listen intently,
gazing up into the canopy. But he had little luck that day. He
brought home one bright green bird the size of a sparrow. One
moonrat, its pointed muzzle messed with blood. Three lengths
of green bamboo to make new blowpipes. And four wild
seedlings of the *petai* tree, whose bitter beans the Semai sell to
the Malays. We had the meat – gamy, like salted lamb – for
dinner, with bowls of dark-green spinachy leaves.

The Semai still live in relative balance with the rainforest.
The only areas they clear are the rice plots, and the settlement
itself – so that storm-toppled trees can't fall on their houses.
There is some erosion before the rice sprouts and matures. But
the forest around catches the lost soil. The rivers are crystal
clear, without a trace of eroded sediment. After the rice harvest,
plots are swiftly recolonized by species from the bordering
forest. Within ten years only a trained eye could spot a former

garden. After twenty years it would take a botanist to tell the difference from pristine jungle.

All around the rainforest stands in its spectacular variety. The ground is matted with star mosses, liverworts branched like channelwrack, purple bellflowers, pink and orange balsams. Treeferns and palms fill the middle storey. Taller trees play host to hundreds of life-forms: lichen, mosses, ladders of climbers, sandy tunnels concealing mountaineering termites. The upper branches are armoured in ferns and orchids.

There is a continual cyclic movement of material. Dead leaves fall to earth in a slow trickling cascade. On the forest floor they lie, mottled with blue-green lichen, perforated with insect holes. The smallest thing that dies nourishes new life.

Like most tribal peoples the Semai treat the forest well. They don't destroy its diversity, to replace it with an impoverished artificial ecosystem of just a handful of crops and domesticated animals. Instead, they use diversity to the full, collecting literally dozens of different kinds of wild roots and leaves, hunting just as wide a range of animals and fish.

The forest repays their kindness. The Semai diet is extremely varied, and the forest provides all the year round. There is no hungry season in the month before the new harvest. They have what seems – apart from illness – to be an easy life.

In most agricultural villages, people are up and about their work even in the half light before dawn. Musoh is the only Third World village I have stayed in where I was usually up before the residents. At eight-thirty, some days at nine, they were still asleep. Normal work took three or four hours a day at most. In the heat of the early afternoons everyone sat around in the small kitchen, with a fire burning on a circular hearth of soil. The men etched their blowpipes or smeared their darts with gorey *ipoh* poison and dried them by the fire. In the evenings the young men played foot-volleyball on the grass, with a light ball of rattan which they kept in the air with their feet.

Such forest peoples tread lightly on the earth. They impoverish no soils, destroy no ecosystems. Their survival does not demand the extinction of any other species. The Semai ask forgiveness of trees and animals before cutting and killing them.

All this is under threat. For the Semai ancestral lands are

surrounded by encroaching Malays and Chinese who are much further down the road of agricultural and industrial 'development'.

A few miles away, over the watershed, lies the foremost tourist hill resort of Malaysia, Cameron Highlands, crowded with hotels, cafes, souvenir shops, golf courses. Shops sell cases of forest butterflies, including the magnificent velvet black and iridescent green Rajah Brooke birdwing. The Semai children are the prime suppliers. Every one of them carries a butterfly net.

Market gardens, growing vegetables for the urban market, are spreading, along with tea estates and, lower down, rubber estates. Loggers are working sections of the forest, and the rivers downstream are coffee-coloured with erosion. There are dozens of small hydroelectric dams, and a high road is planned that will cut right through the highlands.

The Semai of Musoh are adapting to the modernizing economy around them. They have kerosene lamps, pots and pans, blankets. One or two families have bicycles and radios. Musoh even has its first television set, run off an old car battery, though it is more often out of action than in. The men wear boxer shorts and T-shirts, the women blouses and sarongs of Indian cloth.

To buy all these goods, the Semai have to extract more from the forest than they require for their own basic needs. The men and boys cut rattan vines and porter them down to the lowlands, tails trailing like brontosaurus. The women spend most of their spare time cutting up *pandanus* leaves and weaving them into sleeping mats for sale.

And so they are losing their precious leisure, and their priceless culture.

Old Basuloh retains all the Semai mysticism towards nature. Baboi his eldest son, though he has inherited all his father's hunting prowess, already has a more pragmatic view.

Batom, their seventeen-year-old nephew, would make a brilliant hunter. He has all of Baboi's alertness and observational skills. By his age he should know how to make a blowpipe. But he has never learned even how to use one. Indeed he says he has no interest in learning, and would far rather watch his neighbour's battery-powered television or go to discos in the nearest town. He wears flashy trainers, which he bought by catching rare butterflies and leaf insects and selling them to Chinese traders. He knows the price of every species and the value of none.

Within two generations a culture has been destroyed. And the destruction of the culture proceeds hand in hand with the destruction of nature.

Yet we should not imagine that without these outside forces all would be well.

For the most salient fact about the ecology of the Semai is that there are very few of them. They live at very low population densities. Their traditional economy and society are regulated by that central fact.

The Semai are already one step down the road away from nature. The Batek, who live in the remotest forests of Malaysia, are pure hunter-gatherers. They live in temporary huts of leaves. Once their favoured food plants are exhausted in one area, they move camp to another. Among the Semai, population density had passed the limit that can be sustained by pure hunting and gathering: that was why they began to plant rice and cassava.

If their population density increased further, they would have to chop down more forest to grow more crops. Then they would alter their ideas about nature, so as to allow them to alter nature in accordance with their needs. Just like the rest of us did. For once upon a time we all lived like this.

In time, as their numbers increase, their plots will spread, and the forest will dwindle. Then they will be forced to return more and more frequently to their plots, until they are cultivating them permanently. One after another the forest animals, the wild food and medicine plants will die out. The Semai diet will grow poorer, and they will work harder.

The idea of forest peoples exerting no pressure on the environment is mythical. Ever since humans invented fire and sharpened tools, there has never been a time when we exerted no pressure on the environment.

But when population density and consumption are low, and

technology limited to digging sticks and blowpipes, the pressure is slight and sustainable.

Human history is the history of increasing numbers, increasing consumption, and increasingly invasive and disruptive technology. The compound action of these three elements lies behind our increasingly destructive impact on the planet.

1

ONE PART WISDOM:
the great debate

Under totally favourable conditions the power of animals to multiply is spectacular. In 1937 two male and six female ring-necked pheasants were released on Protecton Island, Washington. Within five years they had increased by 166 times, to 1325 – an annual growth rate of 180 per cent.

Within the space of a year field voles can multiply their numbers twenty-four times. Flour beetles can increase 10 billion times. A pair of *Daphnia* water fleas can become 2,000,000,000,000,000,000,000,000,000,000.[1]

The human capacity to reproduce is paltry in comparison. The most children a mother has had in modern times was fifty-five, including five sets of triplets, born to Leontina Albina of Chile over a period of thirty-eight years. Assuming no child deaths, we might in theory be able to double our numbers every three or four years. In practice the highest recorded rates of natural population growth sustained over a decade or more have rarely exceeded 4 per cent.[2]

Shifting into higher gears

For most of human history they have been very much lower than that. Our early growth rates were almost imperceptibly slow, limited by disease, injury, predators, warfare, and the availability of food resources. Around 10,000 BC there may have been no more than 4 million of us.

Most hunter-gatherers number between 0·3 and five persons per square kilometre. The invention of agriculture placed us in control of our food supply. Higher densities – ten to thirty per km[2] – became possible. By the beginning of the present era world population totalled perhaps 170 millon.[3]

Since AD 0 there have been five main phases of growth, like five gears of a car. Each has its own characteristic speed, ground covered, and specific driving cogwheel. In each successive phase up to 1980, the pace of growth increased by three to five times, while the annual addition in numbers multiplied tenfold.

We were in first gear between AD 0 and 800. With economic stagnation, political chaos, and mass migrations, the growth rate was a crawl – 0·03 per cent a year. An average year brought only an extra 63,000 people on earth.

The engine shifted into second gear between 800 and 1700. Gradual agricultural improvements, mainly in Europe and China, were the motive force. The growth rate warmed up to 0·11 per cent, adding an average 677,000 people a year.

Third gear was engaged in the two and a half centuries between 1700 and 1950, powered by the industrial revolution. The growth rate now quickened to 0·57 per cent a year, the annual additions to 7,624,000 extra people. During this phase the human race passed its first billion, around 1820. The second billion, reached in 1930, took just over a century.

We moved into fourth gear around 1950. Two new driving forces brought down death rates in developing countries. One was the gradual introduction of modern preventive and curative medicine, including immunization, improved water and sanitation, and antibiotics. But these alone could not have increased survivals without the second: the agricultural revolution based on chemical fertilizers, irrigation and improved seeds.

The annual growth rate reached an all-time high of 2·05 per cent in the 1960s. An average year saw an extra 64 million people on earth. We reached our third billion in 1960, only thirty years after the second. The fourth billion took only fourteen years.

Since 1980 we have been in fifth gear. The torque is not so high – the growth rate is down, to around 1·74 per cent a year. But we are still cruising along on the momentum of the fourth phase. Because the starting total is higher, the numbers added each year are much higher. Our fifth billion was passed in 1987. It took just thirteen years.

We have not yet reached full throttle. The decade of the 1990s will see the highest annual additions to world population in all history. The 6 billion mark will be passed around 1997, just ten years after we exceeded 5 billion. By AD 2000 the United Nations expects that no less than 969 million people will have been added over the decade. That equals the population of the whole world around 1810.[4]

It is hard to comprehend numbers of this magnitude. The United Nations Population Fund give out a little pocket calculator that helps. Every sixty seconds it updates world population. Take your eye off the display for just one minute and look back: the figure has jumped another 180. Three extra people per second.

An invisible clock is ticking in every country of the world. Useless, these days, for children to learn populations of countries. By the time they're asked again, the answer will be different. The population of India, I learned as a schoolboy in 1960, was 440 million. As I write it is almost double that figure, and is increasing by 19 million a year. When I visited Bangladesh in 1978, the population was 83·5 million. When I went back in 1991, it was 119 million. Dhaka, the capital, had doubled in size.

The 1990s will see two whole Europes, East and West, added in just ten years. An extra United States every two and a half years. Two United Kingdoms every fourteen months. A Sweden or two New Zealands every month. A Birmingham every four days. Every twenty-four hours a town the size of Walsall or Wolverhampton. A school class of thirty every ten seconds.

Such figures are apt to induce panic.

But is the panic justified? Do we need to worry? Is population growth a problem? Is it something we can handle? Or is it, as some would make out, a positive benefit?

Objections to utopia

The Greeks, with a relatively static technology, took it for granted that a city state should balance its population with its resources. Sparta kept the number of her male citizens stable by infanticide. Most Greek city states eased pressure on limited land at home by sending out colonists around the Mediterranean. Plato recommended zero population growth for his utopian Republic. For Aristotle a populous city was hard to govern well. The ideal size of state was one that could be taken in with a single view. Population should be limited by late marriage and exposure of deformed children.[5]

But it was not until the late eighteenth century that the outlines of the modern debate were first traced. Curiously they emerged not from population pressures but from ideological struggles. The population question was just one of the grounds on which egalitarianism and conservatism battled it out. That battle has coloured and clouded the debate ever since.

The terms were set by Robert Wallace in 1761, in his *Various Prospects of Mankind, Nature and Providence*. Wallace recommended equality as a complete remedy for distress and selfishness. But he raised an apparently insuperable objection to his own utopia. It would eventually self-destruct through overpopulation. Children would be so well taken care of that infant mortality would fall and population increase. The earth would at last become overstocked, unable to support its inhabitants. Cruel and unnatural customs would have to be introduced to limit numbers. Women would be cloistered, males castrated at birth. People would be executed when they reached an appointed age. Disputes over these intolerable rules would bring violence and war. Deaths in battle would cull the population to manageable proportions.

Wallace's paradox seemed to rule out any and all egalitarian utopias. Thus the anarchist William Godwin, father of Frankenstein's creator, Mary Shelley, felt he had to answer Wallace's objections. 'The number of inhabitants in a country,' Godwin wrote in 1793, in his *Enquiry Concerning Political Justice*, 'will perhaps never be found, in the ordinary course of affairs, greatly to increase beyond the facility of subsistence.' Even if it did, three-quarters of the globe were still uncultivated, and the earth could support increasing populations for myriads of centuries. It would be foolish, then, 'to conceive discouragement from so distant a contingency'.

In Godwin's paradise on earth, means would be found to extend human life indefinitely. But population increase would not bring this paradise to a close. Our 'eagerness for the gratifications of the senses' would weaken. People would cease to propagate. 'There will be no war, no crimes, no administration of justice, . . . no government. . . . There will be neither disease, anguish, melancholy nor resentment. Every man will seek, with ineffable ardour, the good of all.'[6]

Malthus: the baleful theorem

Godwin's far-fetched dreams provoked a response in the most notorious tract ever written on the subject, Thomas Malthus's *Essay on the Principle of Population*, first published in 1798.[7]

Malthus was as cynical as Godwin was idealistic. By nature human beings were 'inert, sluggish, and averse from labour, unless compelled by necessity'. The passion between the sexes was basic and would never change.

Because of this urge the human population, when unchecked, would always tend to increase in geometrical ratio (1, 2, 4, 8, 16, and so on). But food production could increase only in arithmetical ratio (1, 2, 3, 4, 5). Hence 'the power of population is indefinitely greater than the power in the earth to produce subsistence for man.'

Since food was essential to survival, the effects of these two unequal powers must be kept equal one way or another. Among the lower classes, the sheer difficulty of survival would exert a permanent check. Any excess numbers that were produced would simply die. The lower middle classes – marginal gentlefolk, tradesmen, servants, skilled labourers – might exercise foresight, and marry late or not at all.

Malthus did not propose any measures to rein back population growth. Natural checks worked continually, and efficiently, to keep population growth no higher than the growth in the food supply.

Malthus's baleful theorem was devised not as a sociological or natural law – though it posed as one. Its main purpose was to prove the impossibility of all schemes to improve the lot of workers or to redistribute income. Malthus makes this quite explicit: the theorem was an 'insurmountable difficulty' in the way of the perfectibility of society. It was, he said, 'decisive against the possible existence of a society, all the members of which should live in ease, happiness, and comparative leisure; and feel no anxiety about providing the means of subsistence for themselves and their families.'

The context is significant. The French revolution exploded in 1789, awakening demands for radical reform in Britain. And reform was increasingly needed. Enclosures and the growth of the factory system were depriving many of the rural poor of livelihoods from land or craft. At the same time war hoisted bread prices. In 1795, as an attempt to alleviate poverty, the Speenhamland system was introduced. This supplemented labourers' wages whenever they fell below the bare subsistence level, calculated according to the price of a gallon loaf and the size of the family. Its net effect was to allow employers to cut wages below subsistence level, leaving the poor no better off than before.

Thus Malthus, son of a landowner, wrote from an anxious position of threatened privilege. The first edition of the *Essay*, hastily compiled, poorly documented, loosely argued, was above all a political polemic. The French Revolution was condemned as a 'fermentation of disgusting passions, of fear, cruelty, malice, revenge, ambition, madness and folly as would have disgraced the most savage nation in the most barbarous age'. The Speenhamland system was criticized as encouraging idleness, dissipation, and overproduction of children. Malthus advocated a return to the old punitive type of workhouse, so unpleasant that it would force idlers out to work. Inequality, he asserted, was inevitable: every piece of matter 'must have an upper and an under side, all the particles cannot be in the middle.'

A man of religious bent – he took holy orders in 1797 – Malthus refused to see in the inexorable workings of these 'laws' the sign of a wrathful God. The constant pressure of distress, he wrote, was intended to direct our hopes to the future, and to self-improvement. Evil existed in the world to create not despair but activity. We were not patiently to submit to it, but to exert ourselves to avoid it. Moreover, inequality 'stimulates social sympathy . . . and affords scope for the ample exercise of benevolence'. The poor must be poor or the rich will have no opportunity to practise charity.

Malthus went on to become one of the first anti-Malthusians. In 1803, only five years after the first edition of his *Essay*, he published a second, much expanded, much improved version, which so modified the original argument as to be almost its opposite. The second edition stressed the potential power of self-control among all classes. If this was exercised, population growth need not outrun the increase in food supply. Moral checks could largely supplant natural ones. Indeed, 'an increase in population, when it follows in its natural order, is both a great positive good in itself, and absolutely necessary to a further increase in the annual produce of the land and labour of any country.'

His social model became much more liberal: government should establish national education systems, and involve the lower classes in framing laws. The aim would be to 'approximate them in some degree, to the middle classes of society.'

However, it was the outrageous polemic of the first edition that was remembered. And like all oversimplifications, it fired its opponents and gave them ammunition to launch against it.

Later liberal economists built on his work. Ricardo formulated the 'iron law of wages' – wages can never rise much above or below the minimum level required for the subsistence of workers and the children needed to replace them.[8]

The socialists reply

The first edition of the *Essay on Population* was an onslaught against socialism. Not surprisingly later socialists took up arms against it. The essayist William Hazlitt condemned it as a work 'in which the little, low, rankling malice of a parish beadle, or the overseer of a workhouse is disguised in the garb of philosophy.'[9]

Marx slammed the *Essay* as a 'sensational pamphlet', a 'libel on the human race'. In Marx's view 'overpopulation' was the outcome not of the laws of nature, but of the laws of capitalism. By investing more and more in machinery, capitalism created a surplus army of labourers who could not find employment. It was not true overpopulation, but overpopulation in relation to an economic system. 'The pressure of population,' wrote Marx's patron Engels, 'is not upon the means of subsistence but upon the means of employment: mankind could mutiply more rapidly than is compatible with modern bourgeois society.'[10]

American land reformer Henry George brought the debate much closer to its modern terms in his *Progress and Poverty*, written in 1879 while he was a state gas inspector in San Francisco. George's American background gave a very different perspective. Crowded Europe was already shipping its huddled masses across the Atlantic by the million. America's huge increases in population had been accompanied by huge increases in wealth.

The real cause of poverty, according to George, was not overpopulation at all. It was rather to be sought out in unjust laws, warfare, excessive rents, lack of secure tenancies: all those 'social maladjustments that in the midst of wealth condemn men to want.' Poverty caused population growth, not the other way round, since the poor usually had more children than the rich. India was plagued by famine not because of overpopulation, but because of oppressive government by the Moghuls and the British. The great famine of 1844 in Ireland – often cited by Malthusians as proof of their theories – was in reality the inevitable outcome of landlords' extortion.

These arguments foreshadowed later left-wing views. But George was also a herald of anti-Malthusians on the radical right. Population growth was a catalyst of wealth. Wealth was greatest in those countries where population was densest. Population increase made the members of a society richer, not poorer. Labour was more productive where it worked together with others, through co-operation and specialization. And extra labour added to the productivity of land. 'The increase of man results in the increase of his food.' The Malthusian nightmare would never come about.

The Boserup thesis

One of George's most influential successors is the Danish economist Ester Boserup. The Boserup thesis, as it has become known, is one of the most powerful theories in the history of technology. When it was first launched in 1965, in her seminal book *The Conditions of Agricultural Growth*, the thesis related only to farming. Later it was applied to the industrial revolution. We shall examine the details in Chapter 2.[11]

Malthus speculated how agriculture determined population levels. Boserup turned Malthus on his head. It was population growth that determined agricultural change, she said. And in explaining the exact mechanisms, she put flesh on the bones of George's assertions.

The first farmers were shifting cultivators, clearing forest, growing food for a year or two, then moving on. They might return to an old plot after fifteen to twenty years. But as their numbers increased, they had to return more frequently to the same plot.

Then problems developed. The soil hardened, weeds proliferated, yields declined. Farmers were forced to develop new techniques. Simple digging sticks gave way to hoes, then to ox- and horse-drawn ploughs. To maintain soil fertility they were forced to introduce manure, compost, crop rotations with legumes, irrigation. Irrigation allowed more than one crop to be grown each year.

All these developments kept food production up with population growth. They all involved more labour, and people would invest more labour only when they had no alternative. Population growth provided the compulsion. Without population growth they would not have come about.

Boserup extols the virtues of population growth in occasionally evangelical style. 'Primitive communities with sustained population growth,' she writes, 'have a better chance to get into a process of genuine economic development. . . . A small and stagnant population is unlikely to get beyond the stage of primitive agriculture to a higher level of technique and cultural development.'[12]

The modern Malthusians

It is surprising how little the terms of the debate on population have changed: today's disputes and political alignments echo those of the nineteenth century.

On one side we have the Cassandras. The more extreme blame overpopulation for almost every ill that human flesh is heir to, from disease and poverty, through dictatorship, revolution, war and slow economic growth, to environmental degradation.

Attacking the Cassandras from the left flank are the socialists. For them inequality in all its forms is the disease, and population growth is only a symptom. On the right flank, free-market conservatives see interference in free markets as the only obstacle to everlasting prosperity. They welcome population growth and can see no reason why it should not continue until we have colonized the entire galaxy.

US ecologist Paul Ehrlich has assumed the role of modern Malthus. His 1967 book, *The Population Bomb*, relaunched the controversy in sensational fashion. 'No geological event in a billion years,' he wrote in 1970, 'has posed a threat to terrestrial life comparable to that of human overpopulation.' Ehrlich predicted that sometime between 1970 and 1985 there would be vast famines. 'Hundreds of millions' of people were going to starve to death – that is, unless plague or thermo-nuclear war killed them first.[13]

Ehrlich advocated compulsory measures if voluntary efforts failed. He condemned the giving of aid to 'short-sighted programmes of death control' – in other words, health programmes in the Third World. He took up the infamous 'triage' doctrine. The term is used when doctors classify war-wounded soldiers to prioritize medical care. In the same way nations should be classified and the hopeless cases abandoned to their fate. Ehrlich urged that the United States should halt food-aid shipments to countries such as India 'where dispassionate analysis indicates that the unbalance between food and population is hopeless'.[14]

A more sober and more complex, but still essentially Malthusian approach, came in 1972 with *The Limits to Growth* in 1972, by US economist Dennis Meadows and his team. This Club of Rome Study used a simple computer model to project trends in population, resource use, food production, industrial output and pollution. If business continued as usual, it predicted, there would be a catastrophic collapse of population around the year 2025, due to a dramatic decline in mineral and land resources. This would be followed by a 'dismal, depleted existence' for the survivors. 'Whatever fraction of the human population remained at the end of the process would have very little left with which to build a new society in any form we can now envision.'[15]

Meadows and colleagues tested more optimistic alternatives. But whichever way they turned, growth of population and output always overshot the long-term carrying capacity of the earth, and sudden collapse ensued. If resources were assumed to be unlimited, then massive pollution brought our doom. If pollution was controlled, the crash came when the limits of arable land were reached and food production per person began to decline. If higher food yields were achieved, pollution finished us off a few decades later due to huge rises in industrial output. One way or another cataclysm came, sometime before the end of the twenty-first century.

The limits-to-growth approach was caricatured by critics as a doomsday prediction. But there was one scenario that avoided a catastrophic collapse. This involved massively reduced resource use and pollution per unit of gross national product; reliance on solar energy; a shift out of manufacturing into services; soil conservation; recycling of all wastes including sewage; and stabilization of population at 1970 levels.

Cornucopias and injustices

Extreme positions invited the inevitable reaction. Ehrlich was accused of diverting attention from social reform, from the need to alleviate poverty and curb inequalities in the United States and the world. Critics like US economist Julian Simon detected a whiff of misanthropy in Ehrlich's emotive description of his conversion to the Malthusian cause: Ehrlich 'came to understand the population problem emotionally one stinking hot night in Delhi. People eating, arguing, screaming. People thrusting their hands through the taxi window, begging. People defecating and urinating. People, people, people, people.'[16]

Simon has been the most thoroughgoing critic of the neo-Malthusian position. The anti-people tone of much of Ehrlich's writing clearly offended him, and he responded with a positive pro-life position: pro human life, that is. He endorsed British philosopher Jeremy Bentham's rule for judging actions: the greatest good for the greatest number. Bentham probably meant the greatest happiness for the greatest possible number of the existing community. Simon means the greatest good for the greatest number of human beings possible. The more the merrier. Other things being equal, a greater number of people is a good thing in itself.[17]

The infant Zeus was nursed by the daughters of Melisseus. In gratitude he gave them a horn of the goat-nymph Amaltheia, with the promise that it should always be full of whatever food or drink they might desire. It was the original cornucopia.

Simon's cornucopia is not the result of divine grace, but of human ingenuity. Throughout recorded history the standard of living has continued to rise as human populations have risen. This parallel growth has not come about by chance. The principle cause of increased wealth, says Simon, is population growth. More people mean bigger markets and easier communications. Economies of scale become possible. Productivity improves as larger numbers of factories with higher output learn by each other's mistakes. Above all more people bring more brains to dream up more technical solutions to problems.

Increased population, Simon says, has led to more resources, not less. Almost every significant mineral costs less, in comparison to wages or consumer prices, than it did a hundred years ago. Resources have increased, not depleted, with use. Far from decreasing through erosion, agricultural land has increased and continues to increase. With increased wealth come demands for a cleaner environment, so pollution control measures tighten.[18]

Simon admits that resource shortages can and do occur, but they are never more than temporary. Human inventiveness responds. New methods of extraction are developed, subsitutes that are cheaper and better than the original. We end up better off than before the problem arose. And population growth is the engine that drives us forwards.

Simon carries on Henry George's capitalist arguments. The radical side of George's analysis has its modern followers among left-wing writers like Susan George, Frances Moore Lappé, Piers Blaikie, and the British charity Oxfam. Rapid population growth may not be desirable, but it is only a symptom of other more deep-rooted problems, not a cause.

I shall simplify and summarize the case that underlies individual nuances. The root cause of population growth is poverty. The poor are 'forced' to have large families so that children can bring in wages or care for their parents in old age. Poverty, in turn, is the result of exploitation, expropriation, inequality and injustice. Western colonialism, imperialism, neo-colonialism are all blamed. An unjust economic world order, multinational companies, the growing of cash crops, exploitation of migrant labour – all conspire to keep poor people poor. And they are 'compelled' to overexploit whatever corner of land the wealthy may have left them. Thus inequality and exploitation are the main causes of land degradation and deforestation.[19]

Flanking the socialists are those, like US Barry Commoner, who exonerate population growth and lay the blame at the feet of technology: a wasteful modern technology whose products and waste output have grown less and less degradable, with an ever-increasing impact on the natural environment.[20]

Ideological chaos

Who is right?

It's not easy to see clear. Population has become a battleground on which everyone wields their favourite sword. For free-market conservatives free markets are the answer. Social justice is the solution for egalitarian socialists. More gentle technology and stronger pollution controls for the critics of modern technology.

And there are other armies on the field. Third World nationalists still see concern with population problems as a smokescreen for Western fears of increasing Southern strength – or a new excuse for meddling in their internal affairs. Minority groups cry genocide when there is talk of 'population control'.

Religious groups have weighed into the mêlée. Islam is split over divergent traditions. Orthodox Catholics follow papal opposition to modern contraceptives – based on conceptions of sexuality inherited from the ascetic first centuries of our era.

Anti-abortion stances easily turn anti family planning – despite the fact that family planning is the best way of reducing abortions.

Sexual politics enters the fray. Male supremacy – though not usually explicitly – manoeuvres to deny women the right to determine their own fertility. In Latin countries unfaithful husbands object to contraceptives that might give their wives the power to retaliate.

Unlikely allies fight side by side against family planning: socialists and conservatives, fundamentalist Moslems and Catholics, male chauvinists and feminists horrified by stories of compulsion in China or Bangladesh.

The battlefield has been stamped into a morass. Each ideological position distorts or slants the subject to suit its own needs. Evidence is highly selective, or anecdotal. Debates on population quickly spark into emotion. No topic in the entire field of environment and development excites such deep passions. Population impinges on our deepest attitudes to sex, gender, family, and ethnicity, and connects with our religious and political beliefs. This unique nexus makes it uniquely difficult to see the subject clearly, free of those distorting lenses that our preconceptions and self-interests often impose between our perception and the real world.

Can we reach a more objective view? This book tries. And we shall find that all sides of the serious debate have some part of wisdom, some element of the truth. But only if we move towards a wider synthesis can we get a complete picture.

2

THE O'ERGROWTH OF SOME COMPLEXION:
three billion years of environmental crisis

And God said unto them, Be fruitful and multiply,
and replenish the earth, and subdue it: and have
dominion over the fish of the sea, and over the fowl
of the air, and over every living thing that moveth
upon the earth.

Genesis, i: 28

Environment, says the *Oxford English Dictionary*, means 'the
objects or the region surrounding anything'; or 'the conditions
under which any person or thing lives or is developed.' This
flexible meaning stretches like a piece of chewing gum and
sticks on to almost anything: from home decor to social
ambience, from city streets to forested wilderness.

I shall use environment in a more restricted sense for the
external conditions of life of humans and other living organisms.
External, that is, to each species. For any one species, other
species are part of the environment.

Organisms relate to their environment in three basic ways.
First: the environment is a bank of resources for consumption,
of raw materials needed to maintain existence. That doesn't
mean just food and energy. Even for animals it means territory,
nesting sites, twigs, soil or mud to build homes. Humans
require resources and energy not just to feed and shelter our
biological existence, but to maintain our elaborate and
increasingly massive social and technological structures.

Second: the environment is the site for the physical presence of the animal and its homes, defences and other structures. This presence excludes those of most other species. Humans, once they leave the pure hunter-gathering stage with its temporary shelters, lead the field here in terms of the space taken up per individual.

Third: the environment is also a sink for wastes. Wastes are not only excreta and dead tissue. They include material displaced as a side-product of feeding or home-building. Mole-hills, debris outside badger sets, tree stumps felled by beavers or elephants. Humans are the leaders here too. Our waste and laying waste reaches every corner of the globe.

Yet there is nothing uniquely wicked or idiotic about *Homo sapiens*. We are not the only species that modifies the environment. We are not the only one that manipulates it to its own advantage. We are not even the only organism that creates large-scale environmental change to the detriment of other species.

All organisms impact on their environment through their production and consumption, dwelling places, and waste. They consume what they need for nourishment or nest-building. What they cannot use they excrete or dump. When they die they make the biggest impact of all. Soil humus, peat, coal, limestone, coral reefs and coral islands, all are built of the remains of living organisms.

Often the manipulation of the environment is a deliberate part of the game of survival. When organisms develop the ability to change the environment in a way that helps them to thrive, they leave more offspring and their numbers increase. Evolution helps those that help themselves.[1]

Plants, when they decay, produce acids. These break down rock into the soil which the plants' decendants need to survive. Trees exude chemicals that help rain to condense: they induce their own water supply. Elephants knock trees over and stop forest invading grassland. Beavers rival humans as landscape shapers. They cut trees down, chop them into logs, dig canals to float them along, build dams hundreds of yards long out of hundreds of tons of timber, create miniature lakes covering several acres.

Some organisms have evolved inbuilt mechanisms to adapt their numbers to the resources at their disposal. Territorial behaviour in birds ensures that each breeding pair has a range big enough to provide sufficient food. Where territories cannot be marked out – as at sea or in open grassland – competition for leks or sites in breeding colonies keeps the total numbers using the resource fairly constant. When resources are short, egg production declines in many animals and spontaneous abortions increase.

But many organisms have no such regulatory mechanisms. Some predators eat their prey, and themselves, to the brink of extinction. The old Hudson's Bay company kept a record of trappers' deliveries. Pelt numbers trace the underlying population changes. The graph of lynx and its prey snowshoe hare looks like a double scenic railway. When snowshoe hares had a population explosion, the lynxes' diet improved. More lynxes survived, and a leap in the lynx population followed. The extra lynxes ate up the surplus hares, and more. Then the hare population collapsed. The lynxes starved, and died back to very small numbers. This switchback cycle repeated itself about once every ten years.[2]

In the main it is not the individual species, but the whole ecosystem that prevents one organism from wreaking excessive damage on the environment. Predators limit the numbers of the herbivores they prey on – and vice versa. Herbivores and predators, by eating dominant species, may make room for wider diversity. Competing species limit one another, forcing rivals to focus on slightly different niches. Decomposers ensure that the globe is not buried in excreta and corpses.

Nature is usually protected by its diversity and complexity. This works at the level of the local ecosystem. It may even work at planetary level. James Lovelock, in his Gaia theory, has suggested that life and the planet earth have evolved in tandem. Despite fluctuations in the sun's output, living organisms have kept the climate equable for the past three billion years. Life makes earth habitable for life.[3]

But we can't afford to get too misty-eyed about the goddess Gaia. The mechanisms don't always work smoothly. They don't protect individual forms of life – just Life with a capital L.

And there have been times, long before humans appeared on the scene, when the whole balance of life on earth was threatened by massive environmental crisis.

Revolutions and eco-crises

A crisis is a turning point. Environmental crises present a challenge to the existence of one or more species. They may lead to local or total extinction – or to adaptation, through evolution. Indeed it could be argued that environmental crisis has been a motive force behind many of the major changes in evolution.

Crisis can occur at every level, from the microhabitat of a single organism, to the entire biosphere. It can be caused by factors external to the system, as when a pond dries for lack of rainfall or a massive meteor strikes the earth. Or it can come about internally, through a failure of nature's mutual policing methods. One species or group of organisms succeeds to excess, threatening the basis of its own life and that of others.

Internal crisis comes in two forms, corresponding to two of the ways in which organisms relate to their environments.

The first type we can call a resource crisis: the organism or society eats up the materials or energy sources it needs to maintain and extend itself – to the point where its resources are exhausted and its own survival threatened. The descending trough of every wave in the lynx–snowshoe hare cycle is a miniature resource crisis.

The second type is a pollution crisis. This results not so much from input as from output; less from consumption than from disposal of wastes and side products. Pollution is waste that is harmful to other organisms – or to the waste producer. In a pollution crisis there is no underlying shortage of resources. But the organism's waste output, or the side-effects of its consumption, injure its environment to the point where many other organisms die out. A pollution crisis is initially a crisis for other species.

But it may hit back at the organism that has created it. Eventually the resources on which it depends for its own survival may be damaged. Indirectly a pollution crisis can phase into a resource crisis.

Instinctively we think of environmental crisis as destructive. And destructive it always is, wiping out life-forms, habitats, societies. Gaia is not invulnerable. She suffers from fevers and chills that can make life uncomfortable for whole swathes of species. She has had a number of grave illnesses and accidents. Even Life with a big L has had its close shaves.

Yet crisis can also be creative. Like revolution, it destroys what went before. But it also provides the opening for what follows. An old system dies: a new one is born out of the ruins. The infinite creativity of matter rescues the situation, throwing up new organisms, or new technologies, to control or counterbalance the threat.

Many of the most revolutionary steps in the evolution of life and human society have come about as the result of environmental crisis.

Succeeding to excess

The first crisis in the history of life on earth was a resource crisis. It left no record, and can only be inferred.[4]

Comets, lightning and ultraviolet radiation gradually turned the early ocean into a brew of amino acids. The first self-reproducing organisms probably lived by slurping up this high-protein broth. But they were drawing down capital that had been built up over hundreds of millions of years. At some point the original larder must have grown bare. The first heterotrophs were at risk of eating themselves out of house and home.

As the organic soup grew thin, organisms that could produce their own food would have a big advantage. Several types of bacteria evolved to fit the niche. One of them came to rule the earth in its day. Cyanobacteria, blue-green algae, used photosynthesis to convert the inexhaustible energy of the sun into chemical energy to sustain themselves. Life no longer depended on the depleting amino acid soup: it became self-sustaining.

In time these humble single-celled organisms transformed the earth more radically than the human race has yet managed. The earth's original atmosphere was probably like that of Venus and Mars: more than 95 per cent carbon dioxide, with around 3 per cent nitrogen, and traces of other gases. But no free oxygen at all.[5]

Yet the earth's atmosphere is now totally different: 77 per cent nitrogen, 21 per cent oxygen, 1 per cent water vapour, 1 per cent the inert gas argon, with less than one-two-thousandth of 1 per cent of carbon dioxide. This transformation was largely the work of of the blue-green algae.

Their appearance precipitated a second crisis, this time a pollution crisis. They made their food harmlessly enough, from solar energy, water and carbon dioxide. But it was their waste gas – oxygen – that brought problems. At first it combined with iron in the sea. The ocean rusted, and iron oxides precipitated to the sea-bed as banded ironstone. Then the land rusted. When these sinks could absorb no more, free oxygen began to accumulate in the atmosphere. The ozone layer, which shields life from damaging ultraviolet radiation, was formed. Without it life might never have been able to colonize the land.

But there were minuses. Oxygen has a lecherous appetite for coupling with other elements. Poisonous to organisms that have not learned to harness it, it killed off the dominant bacteria of the previous era, the methanogens, in astronomical numbers. The methane bacteria, giving off the greenhouse gases methane and carbon dioxide, had helped to keep the climate warm.

The surviving methanogens retreated to the margins, into swamps, wetlands, the guts of animals. Methane and carbon dioxide output declined. And the concentration of carbon dioxide in the atmosphere dropped as cyanobacteria consumed it. The atmosphere began to cool. Some 2700 million years ago the earth entered the first and longest ice age, the Huronian, which lasted for 900 million years.

Evolution rode to the rescue. Animals, inhaling oxygen and exhaling carbon dioxide, multiplied and ate the photosynthesizers. Although they were and still are vastly outweighed, they pumped out just enough carbon dioxide to raise the temperature of the earth to equable levels again.

What a piece of work is man

We have not yet had quite as much impact on the globe as the cyanobacteria – but it's clear that we are capable of doing so.

We are not alone in the animal kingdom in possessing technologies. The humblest termite has architectural techniques of moisture and temperature control that we are only beginning to learn. Many birds and mammals use tools, or transmit technologies by learning rather than genetic inheritance. We are not alone in possessing language, or in understanding and investigating cause and effect.

We have an advantageous arrangement of thumb and fingers and a vocal set-up that allows an extraordinary range of sound to be transmitted. But what sets both these capabilities to work is the size of our spare brain capacity – unprogrammed by genetics and therefore free for programming by ourselves.

This does not free us from evolution. We too have to adapt or die when faced with environmental crisis. But now we adapt through cultural, not genetic evolution. Genes change slowly. If the environment changes faster, the organism dies out. Humans can adapt faster still, simply by reprogramming spare brain capacity.

We invent technologies to manipulate the environment. We expand and improve our inventory. We build on what went before. When circumstances change, we change our technologies. We adapt even our own behaviour, liberating ourselves from instinct. In the service of ideas we become capable of things that defy all our instincts: mindless obedience, lifelong sexual abstinence, martyrdom. Our social organization becomes another of our technologies.

These talents have enabled us to adapt to environmental change better than any other organism. Shortage of food has rarely held us back. We are omnivores, and have learned by trial and error to eat thousands of different plants and animals. The invention of fire and clothing liberated us from a limited climatic range. We thrive from the hottest deserts to the coldest Arctic. Predation has not kept us down. Since the first flint was sharpened we have sat, like an Inuit over a hole in the ice, at the top of every food chain. Our only systematic predators are other humans, yet all the murders, raids and wars of history have done nothing to slow our expansion. Indeed they have honed our genetic and cultural abilities to organize, to invent, to control.

Yet we too have faced major environmental crises before the present one. Two of these were of our own making. And they stimulated two major leaps in human technology and society – the agricultural revolution and the industrial revolution.

We survived both, and emerged with even greater powers to escape nature's control.

Who would fardels bear? The agricultural revolution

Until a couple of decades ago hunter-gatherers were viewed through agrocentric eyes. They were thought to lead miserable, brief, animal lives. The invention of agriculture supposedly came as a sudden liberation and was joyously embraced. Agriculture alone permitted rapid population growth, and provided the foundations for civilization.

Over the past few decades these stereotypes have been inverted. Research has shown that hunter-gatherers enjoy a better diet than their agricultural neighbours, for a lot less effort. Like the Semai, they draw on a very wide variety of leafy vegetables, roots, fruits, nuts, seeds, game and fish – the sort of diet the affluent choose when they have the money.

The !Kung live in the Kalahari desert, one of the least hospitable environments on earth. Yet US anthropologist Richard Lee found they had a typical daily intake of 2140 calories – 8 per cent more than the recommended allowance (RDA) for people of their stature. Protein intake was 93 grammes a day, almost 50 per cent more than the RDA. They were quite choosy too. Out of the 300 edible plants and animals known to them, they ate only a quarter. Unlike their agricultural neighbours, they did not appear to suffer from a lean season. There was less kwashiorkor. The very variety of their food sources protected them from famine. If some failed, others would still thrive. And they managed all this on a leisurely fifteen hours' work per week.[6]

Contrast the life of typical agriculturalists. At peak periods they work from dawn till dusk. The grains they subsist on have to be planted, harvested, threshed, ground and cooked. All this for a monotonous diet, a few tasteless staples that prematurely wear their teeth out. And a less reliable subsistence, depending on artificially bred plants, often outside their natural range, liable to failure from drought or pest attacks.

Who would voluntarily give up what has been called the 'original affluent society' of the hunter-gatherers, to grunt and sweat under the weary life of agriculturalists? Most hunter-gatherers are well aware of the principles of planting and tending seeds. They know the places and times under which their favoured foods grow. They often have elaborate conservation measures to avoid exhausting them. They cannot fail to notice seeds growing on their waste heaps and toilet areas.

Hunter-gatherers do not maintain their traditional life-styles out of ignorance, but out of choice.

The transition to agriculture was not a sudden breakthrough. Not a wonderful discovery that eased the way to a better life. Cultures that adopted agriculture did so because they were forced to by a resource crisis of their own making, as increasing populations pressed on shrinking wild food resources.

In the Near East, early hunters got a large proportion of their diet from big game such as elephants, rhino and hippo. When these became extinct, from 18,000 years BC onwards, the diet broadened to include a much wider range of plant and animal foods. This so-called broad-spectrum revolution was the first adaptation to man-made shortage.

Then, from about 10,000 BC onwards, there is evidence of increasing reliance on gathered cereals: storage pits, grinding slabs, sickle blades. Human skulls begin to show severe tooth wear and loss – a symptom of cereal eating.[7]

The fertile crescent, sweeping around Mesopotamia from the hills of Judea to the Zagros, may be desolate now. But 12,000 years ago it deserved its name. Here the wild ancestors of wheat and barley, cattle, goats, sheep and pigs lived in close proximity. In some areas wild einkorn wheat grew in stands so thick that a family using primitive sickles could harvest enough, in just three weeks, to live on for a year.[8]

As long as wild supplies sufficed, there would be little incentive to start planting seed deliberately. But gradually there is increasing interference in the natural habitat. Axes are found – a sure sign that trees are being cut. Fire seems to have been used to keep woodland open to create favourable conditions for pasture and for cereal grasses to grow.

The transition to agriculture happens from around 8000 BC. Agriculture was not, as we used to be taught in school, a one-off invention which spread like wildfire as soon as it had been discovered. It was a slow business stretching over several thousand years. During that time the share of gathered and hunted foodstuffs gradually declined, and that of planted foods grew until it came to dominate the diet.

Plants show growing signs of domestication. In wild cereals, the rachis – a segment of stalk that holds the kernels on – grows brittle as the plant ripens, so it will shatter and spread the seeds. In domesticated wheat the rachis strengthens. Seeds don't scatter to the ground on harvesting. The husk separates from the grain more easily, which facilitates threshing. As humans grow increasingly dependent on cereals, domesticated cereals come to depend on humans for their propagation.

There are several theories as to how population increase led to the gradual spread of agriculture. US archaeologists Philip Smith and T. Cuyler Young have suggested that people congregated around places where wild cereals were locally abundant. Because the harvest was hard to transport, they settled in one place. Sedentarization and plentiful food brought population increase. Childbirth and infant-rearing became easier, and the old and infirm could more easily remain with the group. Eventually populations would grow towards the limit of the carrying capacity of the wild stands. There would be more and more pressure to increase the area of edible grains by deliberate planting.[9]

An alternative theory, advanced by Lewis Binford and Kent Flannery, suggests that agriculture originated not in these areas of abundance, but in neighbouring marginal zones. As people learned to live off a broader spectrum of foods, population increased in the most favoured zones. Small groups of these sedentary gatherers budded off into more marginal areas, already inhabited by more nomadic hunter-gatherers, creating excess population pressure. The migrants were forced to re-create the dense stands of wild cereals that they remembered from the optimal zone. But since cereals were not adapted to the marginal zones, they could only be maintained there artificially.[10]

US historian Nathan Cohen pondered why agriculture was adopted, independently, in four or five separate centres ranging from the Andes and Central America, through China and South East Asia, to the Near East. Why did such a large proportion of the human race shift to agriculture, in a relatively short time-span, between 9000 and 2000 years ago? Cohen proposed that a very slow global population increase had spread humans across the globe. This would have built up pressure gradually across wide areas. In other words, at the time agriculture appeared, the world was already full in relation to the technology of hunting and gathering.[11]

Land degradation may be a fourth factor in creating higher population density. If excess populations from favourable centres spread into marginal areas, these would be degraded by the pressure of harvesting wild cereals and clearing forest to help them spread. Later they would have to be progressively abandoned. People are then forced back into smaller areas, creating excess population pressure. At this point people are forced to turn to agriculture.[12]

Permanent revolution

The agricultural revolution is a permanent revolution: a never-ending adjustment of technologies to the availability of land and the numbers of people to be fed from it. There is a continuous increase in intensity, on a scale that leads from shifting cultivation (like the Semai) to permanent agriculture with two or three crops a year.

Before the emergence of the market, increasing population density is the chief stimulus to agricultural change. As Ester Boserup has shown, population growth stimulates changes in technology to meet the demands of population growth.[13]

In the early days of farming forest or woodland, people are few and land is plentiful. Plots are farmed for a year or two, then left fallow. During the fallow period, microbes and rain carrying dissolved nitrogen build up soil fertility. Tree roots, reaching deep into the subsoil, bring up nutrients that have been leached out of the top layers. When leaves fall, these nutrients are restored to the surface.

Nature does the backbreaking work. All the farmer needs to do is to cut down the trees and burn them. Worms, grubs and other soil creatures make the soil friable. Seeds can be planted with a sharpened stick. The forest shade suppresses low-level vegetation, so there are few weeds.

Up hill and down hill of the plot, forest and thick wild grasses act as fairy godmothers for soil and water conservation, collecting rainwater, filtering it into the soil, preventing erosion and floods, ensuring a constant supply of moisture.

But as population density grows, plots cannot be left fallow as long as before. Everything begins to change. Trees don't get time to regrow. Shrubs and, later, grasses come to dominate the fallow, leaving their seeds and massive root systems to create weed problems for crops. Worms are less numerous. Organic material declines. The soil is exposed to sun and rain for longer, and gets harder to work. Soil fertility is no longer fully restored. Yields start to decline.

Generally farmers do not watch passively as the problems pile up. They adapt. The digging stick gives way to the hoe or mattock to break up the land and cut weeds. But this means more and more labour. Food production per person holds up, but only because each person puts in longer and longer hours. At Gbanga in Liberia, where land is farmed one year out of ten, farmers put in about 770 hours per year on each hectare of land – two hours per day, averaged over a year. But at Bamunka, Cameroon, where land is farmed permanently, they put in 3300 hours a year on each hectare – four times a much.[14]

As farming intensity increases, life gets harder. Just as no one would voluntarily shift from hunting and gathering into farming, so no one would move from long-fallow cultivation to permanent farming, unless they were compelled. And once again, growing population density provides the compulsion.

At some stage the demand for extra labour is more than humans can provide. At this point they make a further adaptation. They turn to draught animals and ploughs. This transition helps with weeding and land preparation. And it deals with the fertility problem as well. The animals' manure is applied to the fields and helps to maintain or increase yields.[15]

Rise, Peter; kill and eat[16]

When human beings began engineering nature, they began engineering their own societies, and their own minds.

The agricultural revolution, as it moved from shifting cultivation to permanent farming, brought massive changes in social structure and culture. Control over nature led to control over people.

Hunter-gatherer bands are anarchical, egalitarian, easygoing. There is a premium on individual resourcefulness in finding game or wild plants. There is rarely a surplus over the needs of subsistence. If one person kills a big animal, everyone in the band gets a share. Not much is needed in the way of political organization. Such authority as does exist is limited to the task in hand, and depends on skill and wisdom. Status is earned by generosity in giving.[17]

The earliest stages of agriculture bring modest changes. At first private property in land is uncommon: there is so much of it that no one needs to claim ownership. The primary group may assert control over a certain area of forest, to keep outsiders out. And as the fallow period shortens, they may invest a chief with powers to allocate land.

Then population grows, and farmers return at shorter intervals to the same field, until they are cultivating it more or less permanently. Everyone's fields meet up, and there is a shortage of land. At this stage a field will generally be handed down from father to son and be recognized as the property of that family. The larger society may recognize this state of affairs in its laws, and private property in land becomes the norm. Permanent claims are staked. Private ownership of land develops.

Then village lands spread until they meet the territory of adjoining villages. Disputes and conflicts arise. Centralized authorities are needed to resolve them – and to organize defence and aggression against neighbours. Urban settlements grow in size and become cities.

Permanent agriculture brings with it hierarchies of wealth, status and power. Surplus grain can be stored. Land can be mortgaged and forfeited, bought and sold. Wealth accumulates in fewer hands. Status increasingly depends on amassing possessions, on hoarding rather than giving. Some people live not by direct cultivation but by directing or exploiting the labour of others. Classes emerge. Warfare spreads, and with it slavery. Male dominance increases. The ultimate consequence of permanent agriculture is the huge empire.

Empire represents the total denial of local environment.

Imperial élites live in capitals far from the periphery where agricultural production takes place. Soils and peoples are plundered in pursuit of tribute and slaves. Crops and trees are deliberately burned to harm the enemy. In times of peace animals are slaughtered for sport or gluttony. The wildlife of the Roman Mediterranean was decimated to fill the amphitheatre or the banquet platter.

Culture changes in parallel. The hunter-gatherer band is, of necessity, in tune with nature. The religion is usually animist. All living things, and many non-living, possess soul and demand respect, just like humans. There are elaborate rules, closely linked to the local environment, controlling which animals may be hunted, and when. Technology is the primary limit on human impact on the environment. But culture reinforces those limits.

Early agrarian states often retained gods and ceremonies linked to the natural forces controlling agriculture and the seasons of the farming year. But as warfare intensified and became the general condition of life, transcendental religions emerged: Zoroastrianism, Buddhism, Christianity and Islam. These religions are transcendental in two senses. They transcend the real present world, teaching that a future invisible world is more lasting and more significant. And they transcend locality – offering universal messsages, designed for a universal audience, divorced from particular environments.[18]

The transcendental religions hastened the death of ecological awareness. The present world became a transit hall to the next, a human threshing ground where grain is divided from worthless chaff. God and soul are separate from matter, not inherent in it.

The agricultural revolution did much more, then, than transform the natural environment. It led to the development of ideologies which legitimated the dominion of men over nature; of men over women; and of men over men. These provided the basis for environmental destruction right up until the present time.

The second environmental crisis and the industrial revolution

The first resource crisis revolved around food. The second was an energy crisis.

The schoolboy's view of the industrial revolution is that it was due to inventions like the steam engine or the spinning jenny. The truth is that these inventions would never have been made were it not for a deepening energy crisis in Europe from the seventeenth century onwards. That crisis was the result of populations pressing on limited supplies of timber and fuel.[19]

The Black Death marked a pause in the growth of Europe's populations. It was almost a century and a half before they regained their 1340 peak of 73 million. For the next two hundred years, growth was slow, averaging less than 0·2 per cent a year, but the needs of growing populations for farming land and wood continued to lay heavy pressure on woodlands and forests.

Wood had a surprisingly wide range of uses. It was the main fuel for homes and the basic construction material for houses, ships, machinery, pit props. Freiberg's silver mines used over 60,000 cubic metres of timber a year for props and fuel. Wood was the principal energy source for industry. Four cubic metres were needed to make a tonne of pig iron, another 9 cubic metres for a tonne of wrought iron. Wood ash was one of the main raw materials in making glass, soap, alum and saltpetre.

From the mid sixteenth century local shortages began to develop. As early as 1560, Slovakian smelting works at Stare Hory and Harmanec were forced to cut production savagely because of scarcity of timber in neighbouring forests. Lack of timber held back shipbuilding in Venice and the Bay of Biscay. By the seventeenth century serious wood shortages had spread to France and England. These began to constrain shipbuilding, smelting, iron production and other industries.[20]

The price of firewood rose tenfold in England between the latter half of the fifteenth century and 1700. Inevitably, people turned to substitutes. In the mid seventeenth century there was 'so great a scarcity of wood throughout the whole kingdom' that people were burning sea-coal or pit-coal. John Stowe, in his chronicle of early seventeenth-century London, complains of fog due to coal-burning. Coal replaced wood fuel in glassmaking, brewing, smelting of lead, tin and copper.

Demand for coal surged. In England, output rose from 200,000 tonnes in the mid sixteenth century to 3 million tonnes in the 1690s. But this demand led to other problems. Coal mines had to be sunk deeper and deeper below water tables. Traditional drainage methods using horse-driven pumps could no longer cope. This challenge spurred the development of the steam engine. 'Technological innovation was more effect than cause,' writes economic historian Samuel Lilley. 'The development of the steam engine came . . . hesitantly and reluctantly, as and when it proved no longer possible to cope with expanding needs by traditional means.'[21]

Unlike wood, which was widely distributed, coal deposits were localized, and coal had to be transported in bulk over long distances. British historian Richard Wilkinson suggests that this provided a powerful stimulus to the development of canals, and then railways. The early chemical industry blossomed around the articifial production of alkalis for making soap, glass, alum and saltpetre, to replace wood ash. The brick industry developed as timber for building grew scarce.[22]

In his 1976 book, *Poverty and Progress*, Wilkinson boldly extended the Boserup thesis. Population growth was not only the driving force behind agricultural change. It explained changes in industrial technology too. Land-based resources such as wood or wool grew scarce due to population pressures on land and forest. The industrial revolution replaced land-based resources with mineral ones. 'Industrial expansion was . . . liberated from the constraints of the land supply.'[23]

'These initial changes were made under duress,' writes Wilkinson. 'They were not introduced when the traditional economic system was functioning in ecological equilibrium, but when scarcity threatened the continuation of the established system.'

The third crisis

Since the end of the last glaciation, the major environmental changes we have had to respond to have been of our own making.

Our hunting and gathering activities continued for perhaps 300,000 years. Then slowly rising human populations depleted the natural food supply. We coped with this first world food crisis by developing agriculture.

Within 10,000 years, Western Europe hit the second crisis, as expanding populations and towns pressed on the wood supply. We weathered this, the first world fuel crisis, by shifting to fossil fuels. The industrial revolution ensued.

Now, within 300 years of the second crisis, we have entered the third.

It will lead either to our destruction, or to a third revolution.

3

BOUNDED IN A NUTSHELL:
the new limits to growth

> The improvements to be made in cultivation, and
> the augmentations the earth is capable of receiving
> in the article of productiveness, cannot, as yet, be
> reduced to any limits of calculation.
> Godwin, William, *Enquiry Concerning Political Justice*,
> 1793.

The third crisis is quite different in nature from the previous
two. Quite different from the one that was expected in the 1960s
and 1970s. At that time many thought that population growth
was pushing humanity towards a resource crisis. That we would
simply run out of the things we consume to live and prosper:
food, fertile land, energy, minerals.

On the eve of the twenty-first century, there seems to be no
impending global shortage of any of these essentials.

Cook, little pot, cook

When Jesus fed five thousand people from five loaves and two
fishes, the leftovers alone filled twelve baskets. The history of
fossil fuels and minerals has been almost as miraculous. The
more we have used of many materials, the more we have had
left – so far.

We can think of the global stock of any particular mineral as
a huge buried mountain. Like an iceberg, only a small part is
easily accessible. This portion is the reserve – that part that can
be exploited using current technologies and yield a profit at
current prices. Below this is the reserve base – a much larger
segment which it may not be economic to exploit until prices

rise. Deeper still – metaphorically – lies the resource: the mass
of the buried mountain that we know or can intelligently guess
about. Much of this is not, at present, economically or
technically exploitable – though it may be one day. Finally, far
out of sight, there may be further stores of which, as yet, we
have no inkling.

Most people think of mineral reserves as a pile that shrinks
the more we pull out of it. But reserves defy all the laws of
arithmetic. Take the curious case of copper. In 1950 copper
reserves amounted to 100 million tonnes. Over the following
thirty years 156 million tonnes were consumed. And at the end
of that period reserves stood at 494 million tonnes – five times
the 1950 level.

Reserves are constantly changing. Technological progress
and price changes alter the share of resources which can be
classed as reserves. Exploration adds to our knowledge of
resources. According to economist William Vogely investment
'creates' reserves out of the stock of resources, in just the same
way as investment creates new manufacturing capacity in
response to price and technology changes.[1]

Hence the enigma of simultaneous depletion and expansion.
Aluminium consumption between 1950 and 1980 would have
almost exhausted the 1950 reserves. Yet reserves in 1980 were
3·7 times bigger than in 1950. Over the same three decades the
starting reserves of lead were used up twice over – leaving
reserves more than three times bigger at the end.[2]

Present reserves of silver, tin, zinc, mercury and lead have
lifetimes of twenty-two years or less at current rates of use. But
most fundamental minerals have a long life expectancy. 1988
reserves or iron ore would last 167 years at current depletion
rates. Aluminium would last 224 years.

Oil reserves have been given varying spans to live: but the
date of their final exhaustion has been postponed again and
again. In 1950 there was enough for only eighteen years. In
1989 they could last forty-four years, at much higher rates of
use. 1989 reserves of natural gas would last fifty-eight years at
current use rates. Coal reserves would not run out for more
than two centuries.[3]

Another sign that there is no immediately impending scarcity
of resources: the real prices of most minerals have been on a

long-term decline. Between 1948 and 1989 the trend of commodity prices fell by almost 45 per cent in relation to manufactured goods. There was a brief surge in the 1970s, but this was linked more to the market muscle of OPEC, the oil exporters' cartel, than to any underlying scarcity.[4]

When scarcities do arise, other adjustment mechanisms come into play. Efforts are made to make the material stretch further. Recycling becomes more attractive. Substitutes are used. Reduction in the end-use is the last resort, and rarely called for.

The oil price rises from 1973 did precipitate brief recessions which reduced end-uses. But this was due to their suddenness, which gave no lead-time to adjust. They were soon surmounted and growth resumed. And they gave a shot in the arm for energy conservation. In 1988 Western economies were using 25 per cent less energy to produce each $1000 of Gross Domestic Product than they had been in 1970.[5]

Other materials that had been the mainstay of economic growth earlier this century were also used more efficiently or sparingly. In 1988, for example, developed countries were using 39 per cent less steel per dollar of GDP than in 1961. In the USA the amount of steel and cement used per person began to fall from about 1970 onwards.[6]

The magic porridge pot that has spewed forth riches in the past may work for us for a few decades more. But it would be imprudent to rely on it ever. The lessons of the past cannot confidently be applied to the future. The studies on changes in resource availability look back to times when we numbered only 2, 3 or 4 billion. When the overwhelming mass of humanity was consuming at very modest rates.

There are no reliable precedents for what lies ahead. We have never been so numerous before. And we have never consumed at such rates before.

Let us construct a nightmare scenario.

Suppose that, some time after 2100, world population levels off at the latest United Nations medium forecast of 11·5 billion. Imagine that by then poverty has been abolished. Assume that the whole world has achieved the American dream and is greedily overconsuming at the rates of the United States in 1988.

If this were to happen, then the 1988 world reserves of

aluminium would be consumed in only eighteen years instead of 224. Copper reserves would disappear in four years instead of forty-one. Zinc reserves would be used up in three years instead of twenty one. All the adjustment mechanisms we know of could not cope with such a rate of consumption. The globe would be laid waste if resources were mined at that rate.[7]

Nor is the long-term fossil-energy scenario much more consoling. In a world of 11·5 billion people, consuming energy at developed-country rates, today's oil reserves would run out in seven years instead of forty-one. Even our current 220 years of coal reserves would be eaten up in thirty-four years.[8]

Such a scenario is no more than a nightmare. It can never come about.

But it does illustrate one thing. A world of 10 to 16 billion people cannot continue to consume resources at current Western levels. Something has to give.

At some point, probably towards the end of the next century, scarcity of minerals could limit either our population, or our economic growth, or both. We could face a conventional resource crisis. If past history is any guide, this crisis need not be followed by collapse, as the Limits-to-Growth scenarios suggest. More likely it would lead – like the agricultural and industrial revolutions – to adaptations, substitutions, in some cases reductions in consumption.

But long before this resource crisis comes upon us we shall have to deal with the present pollution crisis.

Land and food: Malthus was wrong – but for how long?

> Both the jay-hawk and the man eat chickens, but the more jay-hawks the fewer the chickens, while the more men the more chickens.
>
> Henry George, *Poverty and Progress*

One question lies at the core of Malthus's original argument. It remains the focus of all discussions of population and resources. Can world food production expand in line with rising populations?

Over the past two centuries, Malthus's basic theorem has been turned upside down. Human ingenuity has so far been able to increase world food production in line with the increase in human numbers.

Between 1961 and 1989 world cereal production doubled, from 885 to 1865 million tonnes. Over this same period population grew by only two-thirds. As a result, cereal production per person rose from only 288 kg to 358 kg. In the developing countries, too, there was a substantial rise, from 189 kg to 249 kg per person.[9]

Food prices, in real terms, have been on a falling trend. Even in current US dollars, the average wheat price in the years 1985–8 was $121 per tonne – $5 less than 1975–8. Maize in 1985–8 cost $88 a tonne – $19 less than the earlier period.[10]

Nutritional standards have improved dramatically. The average daily energy intake in developing countries rose from 1939 calories in 1961–3 to 2434 in 1986–8. Over the same period, protein consumption rose by 20 per cent. There was a parallel increase in intakes of retinol, B vitamins and vitamin C.[11]

So far so good. But we cannot safely extrapolate the past into the future. There have been food crises in the past. In Africa there are recurrent food crises in the present. No one can guarantee that there will be none in the future.

Looming limits

And the question: Can the world feed itself? doesn't get us very far. If the answer is yes, all that means is that we won't all starve at once.

In practice the world is not one big happy family, and does not go in for feeding itself. There are surpluses in some countries and deficits in others. Average food intakes in 1986–8 ranged from 1604 calories per day in war-torn Mozambique, all the way up to 3901 in carnivorous Belgium.[12]

Those who have money to buy food can always get it. Those who don't starve in the midst of plenty. The reassuring global trends are no comfort to the world's 1100 million poor, nor to the 500 million who cannot afford a minimum diet for good health and normal activity. Famine relief reaches some people in some situations. But it does not touch the bulk of quiet, grinding, daily hunger among the poor. In 1980 there were forty-six countries with average daily intakes below 2300 calories. Intakes in twenty-three of these actually fell during the next eight years.[13]

Even at regional level there are massive disparities.

Asia has performed spectacularly in the past two decades. In terms of food production, the green revolution has worked. Food production per person in China was 94 per cent higher in 1989 than in 1961. In the rest of Asia it was 22 per cent higher.[14]

But the sheer size of the Asian countries biased the global averages upwards. Most other regions of the developing world have not done well at all. In Latin America, cereal production per person fell by 5 per cent from 1970 to 1989. In the Near East it fell by 18 per cent. Africa fared worst, with a 20 per cent decline.[15]

For most individual countries the trends of the recent past are alarming. Between 1978 and 1989 food production lagged behind population growth in no less than sixty-nine out of 102 developing countries for which data are available. In the Near East and North Africa food production per person dropped in ten out of fourteen countries. In Africa, it fell in thirty-five out of forty-one. The Caribbean and Central America performed no better than Africa, with per capita food production falling in fourteen out of sixteen countries.

Only in Asia (thirteen up, six down); and in South America (eight up, four down) did countries with per capita increases predominate.[16]

Malthus's theorem may not be the global constaint that he imagined. But at the level that matters practically – the individual country – food production often does not keep up with population growth over extended periods of decades. During such periods, in our own generation, millions of people have suffered dull daily hunger. And millions of children have died directly or indirectly from malnutrition.

Even the global figures look less assuring when we look in more detail at the trends. The growth in cereal production per

person has steadily slowed down. In the 1960s it was growing at
1·4 per cent a year. During the 1980s, the improvement was
only 0·2 per cent a year. Indeed since 1977 there has been no
visible upward trend underlying the year-to-year fluctuations.
Cereal production per person in 1990 was slightly lower than in
1978.[17]

And there has been a systematic slowdown in the growth of
all those factors that go into agricultural production: land,
fertilizer use, irrigation and livestock.

The closing frontier

Land is basic. In developed countries the total farmed area has
barely changed since 1970. But in developing countries the
cultivated area increased less than half as fast in the 1980s as in
the two previous decades. And because population was growing
faster, there was an even steeper drop in cropland per person,
from 0·33 ha to only 0·21 ha – or a square with sides 45 metres
long. The decline happened in all regions.[18]

Land is also crucial to the provision of livelihoods. Some 1100
million adults still worked on the land in 1990. In developed
countries the arable area per person working in agriculture rose
from 5·4 hectares to a massive 12·5 hectares between 1961 and
1988, due to mechanization and the shift of labour into industry
and services. But in developing countries, the agricultural
workforce continued to grow. Here the average farmer or farm
worker had only 0·78 hectares to provide a livelihood in 1988 –
down from one hectare in 1961.[19]

When Europe overcrowded in the nineteenth century, her
huddled masses could seek their fortunes in the wide open
spaces of the Americas and Australasia, conveniently peopled
by natives mostly at the gathering or shifting cultivation stage.
But the world's population has trebled in the meantime. The
gaps have filled out, and there are few virgin lands where the
huddled masses of the South could find breathing space. The
frontier is shrinking fast.

The bulk of remaining land reserves lie under rainforest.
There's plenty of it. But it is concentrated in a handful of
equatorial countries. And the soils are often fragile, easily

degraded once tree cover is removed. Hot, humid areas crawl
with pests and diseases. It was not by chance that the
rainforests – the notorious 'green hell' of early explorers – were
last to be settled.[20]

South America is the only continent with comfortable elbow
room. In 1988 it was farming only 15 per cent of its cultivable
land. Africa is farming only 22 per cent. But three-quarters of
her reserves are in the centre and south. Out of 37 sub-Saharan
African countries, 18 will be using more than half their
potential farmland by the end of the century. The Sahel and
West Africa are already farming the whole of their suitable
land. Reserves in these two regions, such as they are, are mainly
marginal.[21]

In other regions the frontier is already closed. In 1988 the
Near East and North Africa were farming all the land that was
suitable for agriculture, plus some that was unsuitable. Asia
was farming 95 per cent of her decent farmland in 1988.
Already India, Pakistan, Bangladesh and the two Koreas are
farming more than 97 per cent of the land that is capable of
being farmed.

The inner frontier

Where land is in shorter supply, inputs that can boost the yield
of land become more important. Extension into new areas gives
way to intensification of the old areas. This is the inner frontier
of agricultural growth.

Yields in developing countries are still low. But this is a
double-edged sword. Low yields indicate slow progress. But the
lag offers that much room for future improvement to accommodate growing populations. African cereal yields, for example,
are a quarter of those in the West – and Africa's population is
expected to grow four and a half times before levelling out.
Latin America's and Asia's yields average one-half of those in
the West – and their populations are projected to double before
levelling out. In these regions, then, if Western yields could be
achieved, future populations could in theory be fed with little
increase in the farm area.[22]

But growth in yields cannot be taken for granted. It is a

constant battle – of crop-breeding, fertilizer application, expansion in irrigation, and fighting pests, diseases and erosion. If the yield improvements of 1961–88 can be sustained, Asian developing countries could catch up with present Western yields in thirty years, Latin America in thirty-three and the Near East in forty-two. It would take over 130 years, at the slow rates of the last three decades, for African cereal yields to catch the West.[23]

Yet there has been a worrying slowdown in many of the elements that are needed to achieve higher yields.

World fertilizer use grew by over 9 per cent a year during the 1960s. But by the 1980s it had slowed to less than a third of that rate. Cattle are important for manure and for power. But their numbers have grown only half as fast as human populations. Irrigation increases yields further. In many cases two or even three crops a year can be grown on the same piece of land. Yet here too there has been a marked slowdown. The irrigated area in developing countries expanded by an annual 2·2 per cent in the 1960s. In the 1980s it grew half as fast. The potential for expansion is still very large. But the easiest and cheapest irrigation schemes have already been built. What remains will be increasingly expensive. The economic returns will be less attractive. And the environmental protests will be louder.[24, 25]

The mathematics of Malthus

One pathbreaking study tried to answer the ultimate Malthusian question: just how many people were the lands of developing countries capable of feeding?[26]

The FAO's 1982 study of population-carrying capacity used data on soil, slope and climate from 117 developing countries. These were compared with the requirements of fifteen basic food crops. The crop that would grow best in each area was selected, and the expected yield was converted into calories. Since human dietary requirements are known, it was possible to work out how many people could be fed.

Farming methods have a huge impact on yields, so three levels of inputs were considered. The high level corresponded roughly to Western European farming intensity. It assumed

high levels of fertilizers, pesticides, herbicides, improved seed
varieties, and conservation measures. For each area it used the
crops that gave the highest output. The low level used only
traditional methods and varieties with existing crop patterns.
Intermediate inputs lay half way between the two. At the end of
the 1980s, measured by fertilizer use, Africa is at the low level,
Latin America and the Near East are about two-thirds of the
way towards intermediate. Asia has almost reached the
intermediate level.[27]

The overall results were widely quoted to prove that there
was no Malthusian threat hanging over the developing
countries. And indeed at first reading they seem hugely
optimistic. By AD 2000, even using low inputs, the developing
countries would be able to feed 60 per cent more than their
expected populations from their own lands. With intermediate
inputs they could feed some 13·5 billion people. Using high
inputs a massive 32 billion people could be fed. This is almost
three times the level at which world population is expected to
level out. And it does not count the lands of Asian communist
countries, or of developed countries, which would double the
farmland of the countries studied.

Most of the regional results seemed equally encouraging.
With intermediate inputs, the regions could feed anything from
2·3 times (for Asia) to 13·3 times (for South America) their
expected year 2000 populations. On past form Africa is likely
still to be at low input level, but even then could feed 1·6 times
its year 2000 population. South West Asia would be unable to
feed itself unless it used high inputs.

But it is not regions, but individual countries that must feed
their people, or fail to. And the country results are much less
comforting.

The huge surplus food-producing potential which the study
identified is focused in a few countries. Brazil alone has almost
three-fifths of South America's food-producing potential. Just
five humid countries account for more than half of Africa's
capacity. Short of clearing their rainforests and importing tens
or hundreds of millions of labourers from hungry neighbours,
these countries will never develop their vast potential.

At the other end of the scale, there are sixty-four countries,
with over 1 billion inhabitants, which would be unable to feed

their AD 2000 populations from their own lands if they were using low inputs. Twenty-nine of these are in Africa, and house 60 per cent of the continent's population. Out of sixteen countries in South West Asia, no less than 15 would be critical, along with fourteen out of twenty-one in Central America and the Caribbean. Thirty-six countries in the world, with 485 million people, would remain critical even if they used medium inputs. Of course many of these countries can export other goods and services to pay for food imports. But others – including most of the African ones – cannot.

These findings are worrying enough. But the picture they present is, if anything, hugely over-optimistic. For the study assumed a spartan world where the land grows nothing but staple foods. Rich and poor alike eat the bare minimum for health and work activity. There are no luxuries: no beverages, no fibres, no fruit apart from bananas, no green vegetables. Livestock are banished to land incapable of growing crops. Almost a quarter of the assumed cropland is marginal, with meagre yields that few farmers would find worth the effort. On top of all this the higher input levels involve radical shifts in the mix of crops that people grow and eat.

To allow for all these factors, at least one-third should be deducted from the potential output. When this is done the results are even more alarming. By the end of the century no less than seventy-five of the 117 countries could not feed their populations from their entire land base using low inputs. And forty-three of these still could not manage using intermediate inputs.[28]

Plumbing the depths

We can increase the produce of the land by farming it. But we cannot yet farm the oceans. At sea, humans are like jay-hawks: the more fishermen, the less fish.

More than two-thirds of the earth's surface is covered by ocean – an area of 362 million square kilometres, deeper, in places, than Mount Everest. Yet already we are pressing against its limits. This vastness has become our garden pond.

Since the Second World War the total ocean fish catch has

risen from 20 million tonnes to 85 million tonnes in 1988. Catches grew very rapidly in the 1950s and 1960s, averaging over 6 per cent a year. Plastic nets, mechanical net-hauling, electronic aids to locate fish, all stacked the odds against fish. Freezing at sea and bigger boats widened the range open to fishermen.[29]

The early 1970s saw disaster. The South American anchoveta fishery collapsed. The total world fish catch actually declined in the first half of the decade and recovered only slowly thereafter.

The 1980s saw renewed growth, but at only half the pre-1970 rate. Catches declined only in the heavily fished North Atlantic, and for several of the most popular species found there – especially cod and herring.[30]

But there were clear signs that humans were approaching the limits of the sea's sustainable yield.

The fate of the whales stands as warning. As each species was overfished and declined, fishing effort shifted to another. When catches of humpback whales bottomed in the beginning of this century, whalers shifted their attention to blue whales. Catches of these peaked in the 1930s and fell off. Fin whales were the next target – but these began to decline after the 1950s. Sperm and sei whales took the slack, but then they too peaked in the 1960s, leaving only minke whales to hunt. This lethal game of musical chairs continued until all the big whales were threatened.[31]

This same process is now at work in ocean fisheries as a whole. As the growth in catches slows in traditional areas, fishing boats head for others. As catches of one species fall, attention shifts to others.

The FAO has estimated the total sustainable yield of the ocean's fisheries at roughly 100 million tonnes. The total catch in 1988 was 85 million tonnes – and this does not include an estimated 24 million tonnes from artisanal fisheries. Thus the overall sustainable yield of the world's oceans was probably already passed in 1988.[32]

Almost all stocks of bottom-dwelling fish and crustaceans are already fully exploited or overfished, the FAO reports. Very few untapped resources of conventional fish species remain. There is still scope to reduce waste during and after the catch. Some fish currently processed into meal or oil may in future be sold for direct human consumption. Yet these measures will not do much more than meet increased demand up to the end of the century.[33]

Of course aquaculture – the rearing of fish in ponds – could well expand to meet the rising demand. But it seems likely that future generations face a dwindling supply of ocean fish: less per person, and at much higher prices.

Water, water everywhere – and not a drop to drink

The name Earth is misleading: a visitor from space might well call our planet Water. But this ubiquity is illusory. More than 97 per cent of the total is salt, poisonous to people and crops, corrosive to industry. Of the freshwater, three-quarters is frozen in ice-caps and glaciers. Most of the rest is buried too deep to tap. The annual run-off from rainfall is only 41,000 cubic kilometres. Of that, discounting flood waters that cannot be harnessed and remote waters that cannot be reached, perhaps only 9000 km^3 are reliably available for human use. Of this, we are already using more than a third.[34]

Water is the most abundant of all resources. Yet it is the most likely to impose severe limits on human development in many parts of the world. Water is fundamental to every aspect of human life and economic development. It is crucial for food production – rainfed as well as irrigated. Agriculture took 69 per cent of world water use in 1987. In many developing countries the share was as high as 99 per cent.[35] For industry and energy production, too, it is a crucial input. These consumed 23 per cent of water withdrawals globally. Annual consumption per person increases with industrial development.[36]

And water is essential for human health and amenity. Most diseases in developing countries are passed on directly or indirectly through water, or caused by lack of sufficient water to wash. Home uses accounted for 8 per cent of the 1987 total. Domestic consumption rises with prosperity. As incomes grow, extra taps are added, extra bathrooms, extra washing machines and dishwashers. People with taps in the house use far more than people who have to fetch water in buckets. Typical use per day rises from around 20 litres per person when wells are the source, right up to 170 litres when there are several taps.[37]

In many developed countries water use has stabilized. Indeed in several it has begun to fall. Industry is recycling more water to reduce output of polluted effluents.

But in developing countries, growth in water use is compounded. Intensification of agriculture demands more water per hectare. At the same time growing industries gulp down more and more. And on the home front the spread of piped water and flush toilets doubles and trebles water consumption per person. On top of all this the number of persons continues to grow rapidly.[38]

Countries using less than 10 per cent of their annual renewable water resources do not usually face supply problems. Between 10 and 20 per cent there may be regional difficulties. Over 20 per cent there are serious problems with storage, large-scale transfers, and so on. America uses 19 per cent – and has severe water shortages in the arid west and parts of the urban east. In Britain, which uses 24 per cent, there are persistent regional problems.[39]

Already in the late 1980s many countries were headed for trouble. Out of 113 developing countries, fourteen were using 10–20 per cent of their resources. China and most of South Asia fell into this band. Another twenty-five countries were using more than 20 per cent.[40]

As with food the tightest situation faces the countries of the Middle East and North Africa. Out of twenty countries in the region, eleven are already using more than half their water resources. Libya, and all the Arabian peninsula save Oman, are using more than 100 per cent. They are relying on expensive de-salinization of sea-water, or drawing on underground reserves of fossil water that cannot be replenished. And populations in the region are projected to double, treble or in some cases more than quadruple before stabilizing.[41]

Much of Africa also faces dangerous shortages, according to Swedish hydrologist Malin Falkenmark. In 1982 only six countries, with a total population of 65 million, faced a situation of water stress or scarcity. By the year 2025 the number of countries affected would have risen to 21, with 1100 million people – two-thirds of the population of the whole continent.[42]

Water shortages will mean competition at every level. Local disputes over wells, canal routes, and irrigation schedules, are already common among farmers in South Asia and the Near East.

At international level there are conflicts between nations that share water resources. In 1975 India built the Farakka barrage to store water for irrigation and to ease Calcutta's water shortage. The dry season flow to Bangladesh has declined, and salt water now pushes further inland. Yet future population growth will force India to siphon off more and more water along the full length of the Ganges.

Egypt already uses 97 per cent of her water resources. But she gets almost no rainfall. All but 3 per cent of her water flows down the Nile from outside her borders. As the upstream water needs of Sudan and Ethiopia grow, so will the scope for conflict. Perhaps the most intense competition in the world exists over the Jordan river, which Israel – already using 88 per cent of her water resource – must share with Jordan, using 41 per cent. Water need is a powerful incentive for Israel to retain control over the West Bank and over southern Lebanon, where the river rises.

At national level shortages mean competition between uses. Agriculture, industry and home uses are all essential to development – yet it may not be possible to satisfy all demands. In some countries one sector or other will be held back.

Take Afghanistan. In the late 1980s the country was using 52 per cent of its water resources – almost all of that for agriculture. Afghanistan's 1990 population of 16·6 million is projected to reach 79 million before it levels out. Suppose the country uses water in homes and industry at European levels. Suppose that crop production requires the same amount of water per head of the population as now. Then Afghanistan will be using 154 per cent of its water resources by AD 2025 and no less than 277 per cent by the year AD 2100.[43]

For many Near Eastern and African countries water will set a development ceiling. It will constrain improvements in agriculture, industry, home use, or all three. And water resource development will eat up a rising proportion of national income.

But it is not only arid and semi-arid countries that face problems. In the USA thirsty cities in the dry states are buying water rights from farmers – whose ability to produce food is thereby reduced. Many areas are drawing down their ground-water resources. Water tables in some parts of Tamil Nadu have dropped by 25–30 metres in a decade.

In parts of Northern China they have been falling by up to 1 metre a year. Beijing's water needs have already caused water tables to drop by up to 2 metres a year and dried up a third of the city's wells. One management study suggests that farmers around the Chinese capital could lose 30–40 per cent of their current water supply over the next decade, to meet city needs that will grow by 50 per cent.[44]

Stop, little pot, stop

Let us sum up the resource accounts.

Many individual countries will face resource crises of growing severity. They will come up against limits of land, water, and food. And those limits will cramp their economic growth, their population growth, or both.

But the world as a whole is not facing a straightforward resource crisis. There is, at least for the next decade or two, no underlying global shortage of minerals, energy and land to meet our essential needs and expected numbers.

Like the magic pot of the tale, the world economy has gone on bubbling out porridge for us. The danger is that the town may be swamped by it, and no one can say the words needed to stop it.

For we are facing a crisis. It is broader in extent than those which led to the agricultural and industrial revolutions. Indeed it is the most serious home-made crisis the earth has faced since the cyanobacteria poisoned the atmosphere with oxygen. It affects the entire global environment, at every level from forests, farms and pastures, to rivers and oceans, right through to the atmosphere and climate.

It is not a resource crisis but a pollution crisis, and our solid, liquid and gaseous wastes are the cause of the problem. It is hitting other species first, just as the rise of the cyanobacteria massacred the methanogens. But we cannot rely on evolution, or Gaia, to muster compensating forces. Life may survive in some lowly form. We may not.

For the effects of this crisis will come full circle, through all the other species and systems we damage, right back to us. A resource crisis could develop, not through our direct consumption, but through our waste and wasting.

Deforestation accelerates soil erosion and climate desiccation. Erosion, desertification and salinization cut food production. We may run short of usable land even though there is at present no underlying shortage.

Acid rain kills forests. Ozone depletion may cut food production. And the full consequences of global warming are incalculable.

We may stop short of the brink. But if we do so we face a different kind of resource crisis: one of enforced abstinence.

There is no shortage of productive land to feed and employ expanding populations: but it is almost all under rainforest. If we cut it down we accelerate global warming. We can compensate for land scarcity by using fertilizer: but fears of water pollution will set limits on how much we can use. There is no shortage of fossil fuels of one kind or another. But we dare not burn them all.

It used to be feared that we would run out of non-renewable resources – things like oil, or gold. Yet these, it seems, are the ones we need worry least about. It is the renewables – the ones we thought would last for ever – that are being destroyed at an accelerating rate.

They are all living things, or dynamic parts of living ecosystems. They may be renewable – but at present they are not being renewed. Over the next seven chapters we shall examine the state of these resources – and the reasons behind their degradation.

4

THE FALL OF A SPARROW:
the passing of biological diversity

Wheresoe'er the traveller turns his steps,
He sees the barren wilderness erased,
or disappearing.
> William Wordsworth, *Excursion*, 1814

Evolution has no visible purpose. We may try to fool ourselves that the unspoken aim is an ascent to higher levels of consciousness. But no self-respecting hen will believe us. There are twice as many chickens in the world as people. It could be argued that the domestic fowl has used us for its own advancement.

Yet evolution does seem to have a direction. The history of life has seen an unmistakeable movement towards greater diversity and interdependence.

That diversity is under threat today as never before in human history. And interdependence makes it more vulnerable.

A brief history of diversity

The rise of diversity has not been smooth. There have been stretches of ponderous climb, long plateaux, then terrifying plunges, mass extinctions in which half or more of all species on earth have been wiped out. The demise of the dinosaurs is the best known of these. But there have been four other catastrophic holocausts, and eleven others of major extent, in the last 600 million years. So far each of these descents has been followed by a further rise.

The first explosion of ocean life that left fossil traces occurred

in the Cambrian period, between 600 million and 500 million
years ago. This wave peaked about 450 million years ago, at the
end of the Ordovician period. At that time there were about 500
families of marine animals. By now existing species were
already exploiting most sources of food, leaving little room for
newcomers. Diversity entered a plateau which lasted around
190 million years, punctuated by two short bursts of extinction.[1]

Then, some 250 million years ago, came the biggest mass
extinction of all time. The great Permian extinction wiped out
75–95 per cent of all pre-existing species. The number of
families was halved, to around 200. There are two prime
suspects. A global cooling may have lowered temperatures,
hitting species adapted to warmth. And around this time the
landmasses of the earth assembled into one giant supercontinent known as Pangaea, exposing species to competition they
had never faced before.[2]

The Permian extinction cleared the way for the second great
wave of diversification, almost as steep as its first burgeoning.
Sixty-six million years ago, the number of families reached 600.

Then, at the end of the Cretaceous period, 65 million years
ago, another mass extinction hit. The toll was not so heavy as in
the Permian crisis. Some 60–75 per cent of species were killed
off. But only 12 per cent of families were left with no survivors.
Among them were the dinosaurs. Huge marine reptiles died out
at the same time, as did the last of the ammonites and the flying
pterosaurs.

Controversy still rages over the cause. Most scientists
currently endorse the theory of a massive meteorite impact. A
thin layer of clay divides the Cretaceous period from the
Tertiary that followed. This layer contains proportions of
iridium usually found only in meteorites. Critics of the impact
theory point the finger at a huge series of volcanic eruptions,
lasting more than half a million years, around the time of the
extinction. Volcanoes could have pumped out the excess
iridium.[3]

The rival theories roughly agree on the mechanism of
extinction. Huge clouds of dust were raised. The sun was
blotted out for months. Global temperatures cooled. Photosynthesis slowed or stopped, cutting food supplies. Acid rain killed
off many types of animals. It also killed off calcareous plankton,

which normally remove carbon from the global cycle. Gradually carbon dioxide built up in the atmosphere, and temperatures rose again.

Death and renewal

All the suspects in mass extinction – meteors, volcanoes, continental drift and climate change – caused global environmental crises.

These crises provided a massive challenge which gave a major boost to evolution. Species that could not adapt fast enough died out. Those that could adapt – or that were fortunately pre-adapted – survived. And they found vast ranges of resources left without species to exploit them. Life abhors a vacuum. Just as a pollarded willow sprouts dozens of new shoots, the survivors spread out into the vacant spaces, in great burgeonings of new species.[4]

The reptiles filled a wide range of niches, from small egg-stealers and flying pterosaurs to lumbering herbivores, giant carnivores and huge marine reptiles. Their demise opened up the field. Mammals and birds diversified to take over all the ways of life that were left empty. Within 10 million years of the extinction there were ancestral bats, whales, horses.

Diversity increases in more positive ways, too. There are crucial evolutionary breakthroughs, opening up whole new fields of resources that could not be exploited before. At each stage there is a critical bound, the invention of a new device, a new meal ticket – a biological parallel to advances in human technology.

Adaptations to breathing oxygen, developed to survive desiccation in drying pools, allowed the vertebrate invasion of land. The hard-shelled egg freed reptile reproduction from dependence on water. The emergence of the wing opened up the unexploited niche of catching insects in mid-air. Mammalian fur and warm blood unlocked the hunting ground of the night in colder climes.

Plants made their own great leaps forward. Roots and vascular systems, circulating nutrients and water, permitted life on land and progressive increase in size. The invention of seeds, enclosed in a case with enough food for a start in life, allowed plant embryos to lie dormant until the right conditions emerged. Flowering plants, with their blossoms, nectars and fruits, no longer depended on wind and gravity for pollination and seed dispersal. They could use insects and animals to spread much further afield.

Each new 'invention' enhanced diversity, added a new layer to the hierarchy of life, filled out the ecological landscape, allowed a new resource to be exploited.

And along with diversity, interdependence grew. As soon as the primordial organic soup began to thin, life itself became a resource for life. Parasites and predators ate other organisms. Decomposers lived off their remains. And more positive relationships developed for mutual benefit.

Interdependence now operates at every level from the regulation of climate through levels of greenhouse gases, down to the microbes in the gut of a termite. Flowers have ant protectors, moth and bat pollinators, bird and mammal seed dispersers – in many cases so closely co-evolved that one species relies absolutely on the other.[5]

Like marriage, such interdependence brings benefits. If one partner flourishes, so does the other. But it harbours corresponding risks. If one partner dies out, so does the other.

How much diversity?

If variety is the spice of life, then life was probably at its spiciest in the interval between the last two glaciations, while humans numbers were modest. There is still an extraordinary diversity. Samples of half a million kinds of insect fill six storeys at London's Natural History Museum. A recent count identified some 1,435,662 species of all kinds that had been named and described.[6]

But the exactness of this figure is misleading. There is no central catalogue, universally agreed and double-checked. Many species may have been counted twice, under different names, by different researchers unaware of each other's work.

Individual variation in some species is very wide. Two samples of one creature might be classed as two species. Conversely, several almost indistinguishable species might be lumped together under one name. The true acid test of separateness – whether interbreeding is possible – is rarely applied. Taxonomists' specimens are usually dead.

All that is certain is our uncertainty. We do not know, to within the nearest half million, how many truly separate species have already been described. We do not know, to within the nearest 20 million, how many species there really are on earth. The smaller and less accessible, the less our degree of knowledge. The rainforest canopy probably hides more species never yet seen or described than the total known to us so far.

Estimates of the number of species range from a conservative 4 million up to 30 million and beyond. But the impenetrable vastness of the forest means that these are very rough guesses based on far-flung extrapolations. The widely cited figure of 30 million, from Washington entomologist Terry Erwin, is based on a sample of just 3099 beetles, knocked down by blasting the canopy with insecticide fog, from nineteen trees in a small rainforest plot in Panama.[7]

Erwin assumed that 13 per cent of the 1200 beetle species he found were specific to the tree they were found on, and that beetles made up 40 per cent of arthropod species. If the beetles were less specialized than that, then the global total of species could be as low as 5 million. If other arthropods are more numerous, the total might range up to 80 million. The higher the figure, the higher the percentage that is made up of insects.

Even the known statistics of bio-diversity are humbling. The animals we most love are the most insignificant. We live on a planet numerically dominated by insects and plants. There are three-quarters of a million named insect species and a quarter of a million plants. All vertebrates lumped together account for only 42,500 species, a mere 1·7 per cent of the recorded total. There are more species of ants than of birds. More of flatworms than of reptiles. More of sponges than of mammals.[8]

Biological diversity is not evenly distributed. It rises with temperature and humidity. It increases as you proceed from the poles towards the Equator, though it is poorer in the dryer regions. Temperate countries have an average of 39 mammal species each, tropical countries twice as many. Greenland is home to 56 species of breeding birds, New York to 105. Guatemala houses 469, Colombia a startling 1395. The oceans follow the same pattern. The Indo-West Pacific harbours five times as many fish species as the Eastern Atlantic, and twelve times as many molluscs.[9]

Why is diversity so much higher in the moist tropics? Higher temperatures may speed up the rate at which mutations produce new species. Temperatures are more even. Plant growth is greater and more reliable, so animals can afford to specialize more. More specialization means less competition: more living room for more species.

Bio-diversity is also higher in areas with a greater layering of the habitat which provides a variety of niches. So at sea coral reefs are the richest in species. The Great Barrier Reef alone supports over 300 species of coral and 1500 of fish, along with 4000 species of mollusc.

On land the tropical rainforest is the most diverse. Its leaf litter, ground herbs, shrubs, middle level and canopy offer a five-storey habitat. Each storey has multiple rooms: climbers, stranglers, clinging ferns, orchids, bromeliads and others. Each tree and plant has its personal range of insect lodgers, each insect its own minute parasites. Each storey and room has its distinct vertebrate predators, with different sets for day and night shifts.

Though it covers only 7 per cent of the world's surface area, the rainforest houses half or more of the world's species. The diversity at local level is astonishing. Where a temperate forest will have twenty-one to forty-eight woody species in a patch of 0·1 hectares, a wet tropical forest can harbour anything from 151 up to 236. Harvard zoologist E.O. Wilson once identified, on a single tree in Peru, forty-three separate species of ant belonging to twenty-six genera – about equal to the entire ant fauna of the British Isles.[10]

By an unfortunate quirk of geography, the way in which bio-diversity is distributed works against it. The tropics have the largest share of the earth's biological wealth. But they also have the fastest rates of human population growth, the greatest need to clear forest and drain wetlands, the heaviest pressures on coral reefs.

The takeover of the biosphere

It has been estimated that some 99 per cent of the species that have ever existed since life began are extinct.[11]

Even when there were no great catastrophes, new species were born and old ones died. It happened at times when the local environment was changing faster than species could evolve, or when species were faced with superior competitors. The modern tadpole shrimp is a positive Methuselah, identical to fossils 180 million years old. The average ammonite species lasted 1 or 2 million years. A typical mammal species in the Cenozoic era lived for a million years. With this average lifespan, of the 4000 species of mammal we know, one would become extinct every 250 years.[12]

The arrival of *Homo sapiens* changed everything. The human race may consider itself the pinnacle of creation, but its most characteristic biological feature is that it devastates creation. We are inimical to all species save our domesticates, and the wily scavengers and pests who still outwit us. The image of the primitive hunter-gatherer living in perfect harmony with nature is a myth. Humans start to manipulate, mould or destroy nature to their own advantage just as soon as they have the technology to do so.

Before we had sharpened weapons, there were giant elephants, lions and kangaroos, bisons with six-foot horn spans, ground sloths as tall as a bungalow, beavers as big as bears. Then came a great wave of extinctions. In North America alone thirty-nine genera disappeared within the space of 3000 years after the last ice sheet melted – twice as many as disappeared in the previous 2 million years. They included two types each of sabertooth cat, of bear and of mammoth, three of tapir, four of camel, six of large ground sloth, seven of deer, moose and antelope, eight of cattle and goats, ten of horse, along with a lion, a cheetah, a wolf, and a giant condor.

The finger points to humans as the culprits. No previous glaciation had ended with such heavy extinctions. The victims were mainly large herbivores and their predators. The plants that the herbivores fed on did not die off. The coincidence with the spread of humans in North America is striking. In Australia and in Madagascar, a surge of extinctions followed the arrival of humans.[13]

The mastery of fire was another massive step in human impact. In dryland areas, camp fires increased the risk of accidental conflagrations. And humans used large-scale fire deliberately, to drive out game, or to bring a flush of lush green shoots to dried out pastures. Repeated fire transforms an entire landscape. It kills most tree seedlings, prevents the establishment of forests, and favours the spread of savannah. Fire-resistant species of trees, perennial grasses with deep root systems, grazers and their predators come to predominate.

The massacre has continued ever since.

Since the beginning of the seventeenth century some 724 extinctions have been officialy recorded. More than half of these have been plants. But proportionately, mammals and birds have suffered more. Out of a total of 9000 recorded bird species, 113, or 1·3 per cent, have already become extinct. Some eighty-three mammals, 2·1 per cent of the 4000 recorded species, are no longer with us.[14]

No two species can live side by side from exactly the same resource. This constraint has propelled diversification throughout earth's history. It has forced animals to diverge from their competitors, to move into new niches, to take up slightly different ways of life.

But we humans occupy every niche that might be of use to us. We move in, exploit it to the full, multiply our numbers, spread into new niches, multiply our wants, expand across the globe. Agriculture was a quantum leap in our attack on other species. We felled trees, moved crops and livestock in, and reshaped the environment to suit them. We marginalized or destroyed their predators, competitors and pests. The industrial revolution further boosted the assault with an endless range of new pollutants and poisons.

The human takeover of the biosphere has reached extraordinary proportions. Stanford ecologist Paul Ehrlich and colleagues have estimated that humanity uses directly around 4 per cent of the total solar energy that the growth of green plants binds. Indirectly we use or waste far more. There are the plant parts we don't eat. The plants our livestock eat, not to make milk or meat, but just to keep alive. The plants we burn in fires to clear the land. And by converting forest to cropland or tarmac, we cut the land's productivity below what it would otherwise have been. Adding all these impacts together, our activities may control up to 40 per cent of net terrestrial plant production.[15]

Even in Pleistocene times, armed only with stone weapons, we managed to wipe out about one species every thirty years – six times the normal pace of extinctions. Our record of destruction accelerated as our technological capacities increased. Between 1600 and 1800, one species of bird or mammal died out, on average, every 5·7 years. The speed of extinction almost doubled over the first half of the nineteenth century, and doubled again in the second half. In the first five decades of the twentieth century, one bird or mammal species became extinct every 1·1 years – 230 times faster than the pre-human rate.[16]

The coming hecatomb

Future historians – if there are any – may well refer to our age as the time of the Holocene mass extinction. It will almost certainly be graver than that of Pleistocene times. Whether it will eventually rival the greatest crises of the past depends on the degree of climate change.

When you look at recorded extinctions there is little sign of this holocaust in progress. Puzzlingly, the number has fallen. Only thirteen birds and mammals are known to have become extinct since 1950, against forty-seven in the previous fifty years.[17]

Some experts put this down to the rise of ecological awareness, and the massive increase in national parks and wildlife reserves.[18]

More probably, it is a statistical quirk. It takes a decade or more before a suspected extinction can be verified. More recent disappearances have not yet been clocked as extinctions. For developed countries, or big creatures, the records probably match the reality. For developing countries, or little-seen species like fish or insects, recorded extinctions are almost certainly just a tiny sample of the total.

Insect species, for example, outnumber mammals by almost two hundred to one. Many are highly specialized and therefore vulnerable. Yet only thirty-four insect extinctions have been recorded since 1600. All but five of these occurred in the United States. More than twice as many mammal species are known to have died out. Such distortions express the distribution of science manpower, rather than the real pattern of extinction.[19]

Most extinctions are invisible: many a bloom dies unseen. A single documented sighting can establish that a species exists. No number of non-sightings can prove conclusively that it does not exist. Belief in the Loch Ness monster is tenacious.

Yet modern expeditions often fail to find a high proportion of species found on older forays. Recent surveys did not turn up a single specimen of more than half the 266 species of freshwater fish known in the Malay peninsula. In Singapore, one-third of fifty-three species of freshwater fish collected in 1934 could not be located thirty years later, despite exhaustive searches. Many of these absences are probably extinctions, though none has yet been counted as such. The usual practice is to assume a species is extant unless it is proven extinct. Perhaps we should reverse the burden of proof, as biologist J.M. Diamond argues, and assume a species is extinct unless proven extant. Or perhaps we need a new category of 'missing species', which are believed, though not known, to be extinct.[20]

All the processes that have led to extinction in the past are more intense now. Despite the lack of recorded data, we can be certain that extinctions are proceeding faster than at any time in recorded history.

How fast? How many species will be lost? Since we don't know how many species there are on earth, or how many are currently becoming extinct, we can only guess. The guesses are little more than intuition, and they diverge wildly. Paul Ehrlich predicted in 1981, for example, that 50 per cent of species would be wiped out by the year 2000 – almost a hundred per day of known species alone. The 1980 *Global 2000 Report* expected losses of 15–20 per cent by 2000. More soberly, the World Resources Institute predicts that 5–10 per cent of plants and 2–5 per cent of birds could become extinct by 2020. Eminent US entomologist E.O. Wilson suggests losses may be occurring at the rate of 2 per cent a decade.[21]

The list of species in danger lengthens every year. In 1989 they totalled 21,806. And the most loved species are in greatest peril. One in every thirteen species of plants are threatened. Of mammals and birds, one in ten.[22]

Some kinds of species are more vulnerable. Big animals that require large territories, or that are attractive game for hunters. Predators high in the food chain. Highly specialized species whose food source is not widely distributed. Species with small ranges or populations. Species that reproduce slowly.

Certain types of habitat are more at risk. Freshwater animals are often found only in individual streams or lakes, and are very susceptible to loss of habitat or exotic introductions. Wetlands are severely threatened by drainage for agriculture and coastal development. Islands are sensitive, because their life-forms have developed in isolation. On Rodrigues Island, in the Indian Ocean, ten of the forty-eight native plants are extinct and another twenty in serious danger. Some 59 per cent of flower species in the Galapagos islands are threatened. So are 66 per cent of Madeira's, and 75 per cent on the Canary Islands.[23]

Causes of species loss

Most extinctions in historic times have been unforeseen and unintended. The introduction of exotics in the human baggage train is often enough to devastate the fauna of islands. The entire surviving population of the Polynesian rat – confined to a single island off New Zealand – was wiped out in 1894 by Tibbles, the lighthouse keeper's cat.

Introduced species spread like disease bacilli, debilitating ecosystems which have developed no immune response to them. They threaten some 19 per cent of currently endangered species. But they have contributed the largest share of past extinctions of birds, reptiles, fish and molluscs.[24]

One of the most devastating extinctions of modern times occurred in Lake Victoria in East Africa. The lake was the only home for as many as 300 species of cichlid fish, with a bewildering variety of feeding habits and mouthparts to suit. In the early 1960s, in the hope of improving fisheries, the Kenyan and Ugandan governments introduced the Nile perch, a voracious carnivore that can grow to the size and weight of man. As the perch multiplied, the cichlids dwindled. The only survivors were those that fed in shallow water where the perch did not penetrate. By 1990 almost 200 of the cichlid species were believed extinct. Typically, none of these has yet found its way into the official extinction lists. If all of them were to, it would instantly multiply the total of fish extinctions by ten.[25]

Human population growth, consumption, and technology are inextricably linked in species losses, but sometimes one or other element dominates.

Technology plays a role. Many coral-reef fishermen use methods that undermine their own livelihoods. Some fish with dynamite, which kills every living thing over an area of 10–100 square metres, and pulverizes the coral. Dynamite fishing has reduced six of Tanzania's eight finest reefs to rubble. Another common scattershot method is to spread bleach, DDT or lindane to windward, and wait on the leeward side to pick up the corpses that float by.[26]

In some places coral itself is the target. In Sri Lanka 40 per cent of the cement industry's materials come from corals and coral sand – a technical choice, based on accessibility and cost, which threatens the reef. Some 75,000 tonnes a year are taken from one six-mile stretch at Hikkaduwa. The unprotected beach has moved 300 metres inland.[27]

The technologies of over-intensive agriculture, encouraged by excess farm subsidies, have damaged many European species. Fertilizer and effluents from livestock lots have polluted waterways. Wild flowers and animals have been poisoned by herbicides and pesticides. In Niedersachsen, Germany, fourteen plant species became locally extinct in the eighty years up to 1950 – but in the following twenty-eight years no less than 131 species were lost.[28]

Some threats to species are mainly the result of affluent consumption patterns. Demand for fur, turtle shells and flesh (for soups), spectacular tropical birds and butterflies, orchids and other rare plants, stimulates local collectors. Where remaining populations of the species are small, this collecting pressure can be a severe threat.

Demand for ivory and horn is the chief danger to elephant and rhino. Numbers of elephants in Africa plummeted from

1,340,000 in 1976 to 740,000 only eleven years later. Older males and females – mature animals with larger tusks – have virtually disappeared.[29]

Affluent demand has a perverse effect on hunting. The scarcer a prized species becomes, the higher the price rises – so extra effort is put into hunting down the last few. Rhino horn imparts a barely distinguishable swagger to Yemeni daggers. For impotent Chinese it is possibly the most expensive placebo in the world. With a single horn fetching up to $44,500 in 1987, almost any degree of hunting effort is worth while. Numbers of the five species of rhino have dropped by 84 per cent since 1970. The Javan rhino is down to its last fifty-five – well below the minimum of 500 or so needed to ensure survival.

Direct predation by humans – whether for food, sport or trade – has been involved in up to 32 per cent of vertebrate extinctions and is a threat to 37 per cent of currently endangered vertebrates. But it's not always easy to distinguish hunting for trade from hunting for subsistence. Many trappers and catchers of trade species are poor rural people seeking a living. Other rural people supplement meagre diets with hunted game. Huge bush rats and snake meat are delicacies in Nigeria. The greater the rural population, the more hunters and trappers there are, the greater the pressure. [30]

No hiding place

The major impact of growing populations is not direct, through predation, but indirect. Loss of habitat to humans menaces two-thirds of threatened vertebrate species.

In developed countries habitat loss is most often due to technology or development for industry, housing or tourism. In Europe hedgerows have been uprooted to make bigger fields for tractors and combine harvesters. In one part of Germany, back in 1877, there were 133 metres of hedge for every hectare. In 1954 there were still 94 metres. Twenty-five years later there were only 29 metres of hedge per hectare. In England and Wales the loss of habitat has been devastating: 95 per cent of traditional lowland meadows, 80 per cent of chalk downlands, 60 per cent of lowland bog, and 100,000 miles of hedgerow have been lost since the Second World War. Lowland marsheṣ, limestome pavements, ancient woodland, lowland heath and upland grassland have all lost between 30 and 50 per cent.[31]

In developing countries habitat loss is more often due to rapid population growth and the resulting need for farm and urban land. We shall be looking at the role of population growth in deforestation and desertification in chapters six to ten.

Africa and Asia have lost two-thirds of their original wildlife habitat. Losses of 80 per cent or more have occurred in Burkina Faso, Burundi, Gambia, Ghana, Liberia, Mauritania, Rwanda, Senegal, Sierra Leone, Bangladesh, India, Sri Lanka and Vietnam. Lowland forests in peninsular Malaysia, the Côte d'Ivoire and Madagascar have been almost totally cleared. In others they have been reduced in size. Asian primates have lost an average of two-thirds of their original range.[32]

Habitat loss tends to be highest where population density is highest. The top 20 per cent of countries, in terms of loss of original wildlife habitat, had lost an average of 85 per cent. Their average population density was 192 persons per square kilometre. The bottom 20 per cent had lost only 42 per cent of their wildlife habitat – and their population density was only twenty-eight persons per square kilometre.[33]

Saved by the bell

Nowhere in the world is population pressure on habitat heavier than in Rwanda.

In the Virunga range of volcanoes, in the north of the country, it often rains several times a day. The young mountain gorillas huddle together in showers. Rain drips through the canopy, trickles off their eyebrows, stands in droplets on their thick black fur. The huge silver-backed male squats, arms wrapped around his waist, half hidden under foliage.

When the rain eases they resume their feeding. The male gets up and moves off up hill. The mothers pull down bamboo shoots while their infants gnaw at the base, bums in the air. Every now and then there is a long, resounding fart. Suddenly the silver-back charges downhill like a runaway tank, tearing down bamboo stalks on the way, straight at the nervous little group of watching humans. But it's his wives he's after. He grabs one, but she's not receptive. She coughs and screeches till he swaggers away, sulking, dragging down another stem to show how strong he is.

The humans watch in awe. We kneel in homage to placate him, pretend to break off seedlings to eat. The guide makes little grunting noises deep in his throat. We have to convince him that we're not a danger to his family.

The area around the Virungas is the only place where mountain gorillas are found. It was declared a national park way back in 1926 – the very first in Africa. At that time the human population of the surrounding Ruhengeri region was probably less than 200,000.[34]

By 1986 the population had trebled to 619,000. The average woman was having almost ten live births over the course of her life. There were 418 people per square kilometre – not far short of the Netherlands, highest in Europe – and rising fast. Land is in such short supply that slopes of 70° are cultivated, so steep it is hard to see how crops cling on.

As population grew, so did pressure on the park. Three times, between 1958 and 1979, large tracts were handed over to local farmers. The original park area of 328 square kilometres dwindled by more than half, to 150 km². Most of the diverse montane forest was lost.

There remained a narrow band around the base of the volcanoes. Bamboo forest graded into open woodland. Then subalpine giant heathers and lobelia rose to lichens and mosses near the peaks. On this tiny remnant human population pressure continued to grow. Surrounding farmers would go in to graze cattle, cut firewood and bamboo, collect honey from wild hives, hunt antelopes.

The impact on wildlife was harsh. Many carnivores disappeared. Lions were last seen in 1943. Several of the 180 species of bird were highly endangered. In 1960 US naturalist George Schaller estimated that there were 400 to 500 mountain gorillas. A census in 1976–8 found only 268. More worrying still for the future, the gorilla population was ageing: only 36 per cent were immature. But as their scarcity increased, so did their value, and the gorillas themselves became prey. Hunters would kill parents to take the young to dealers, for sale to Western fanciers and zoos. In 1978 alone eight were killed.

If things had been left to take their course, the gorillas, and many other species in the park, would probably have died out. The sheer pressure of population would have been the chief cause.

But in 1979 the Mountain Gorilla Project, stimulated by the threat, began. Guards were employed and armed. Protective barriers were put up around the perimeters. Guides were trained to show tourists around. An education project taught local schoolchildren and adults about the value of the park forest in protecting water sources and soils. Gorilla numbers began to recover – a 1986 census counted 294, and almost half of them were immature.

Population growth still preserves the tension between the park and the local people. Previously the park bore the costs of population expansion. Now the local people bear the costs of the park's presence and of continuing population growth. Seven out of ten Ruhengeri farmers consider their land insufficient to meet even their present needs. Soil erosion is heavy and averages 26 tonnes per hectare each year. Three out of four farmers believe their fields have grown less productive over recent years.

Danial Ntawuhora was given a two-hectare plot in 1968. He has since had six children. Two of his sons are married, and he has had to give them their own plots carved out of his. They have only one-eighth of a hectare each. Daniel once kept cattle, but has had to sell them since he can't graze them in the forest, nor can his land grow enough fodder. Buffalo burst out of the park, trample and eat his crops. Now humans are a threatened species in the area.

Loss of species can be mathematically related to habitat loss.

The number of species in a given area is related to its size. The smaller the area, the fewer the species. This is because a smaller area is less diverse and has fewer niches. It also harbours smaller populations of each species, and smaller populations are more vulnerable to disasters or random fluctuations.

Species numbers decline more slowly than the area of the

habitat. When pieces of land are cut off to form new islands, in dams, for example, many bird species become locally extinct within a century. On islands of 25 square kilometres the loss to extinction is 10 per cent. On islands of only 1 square kilometre the loss is 50 per cent.[35]

The decline of forest species may be faster, for a given decline in area, than on islands. For the world's rainforests are not shrinking all in one piece, like balloons with a slow puncture. Rather they are being eaten away from without and within, like a seasonal pond that dries to a few muddy puddles. Deforestation creates land islands, separated by inhospitable arable or pasture which species cannot cross. The smaller the patch, the fewer the species that remain in it. One patch of 14 square kilometres in Brazil, cut off for a hundred years by agricultural clearing, lost 14 per cent of its species. Another patch of only 0·2 square kilometres lost 62 per cent.[36]

Some forest species are very localized, and their home may be the particular tract of forest that is cut down. Terry Erwin's sample of 3099 beetles belonged to no less than 1200 separate species – only three individuals per species in an area of 112 square metres. In the western Amazon basin, two 12-metre-square plots, a mere 50 metres apart, had only 9 per cent of their sampled beetle species in common.[37]

Large herbivores and predators are more vulnerable for the opposite reason. They need larger territories to feed them. Once a stretch of forest falls below the minimum area for a viable breeding population, extinction is only a matter of time. The death of these larger species has a domino effect on other species, since they structure the whole ecosystem. Large herbivores are the only animals capable of dispersing some big tree seeds. They may also keep dominant plants down, leaving room for other species. Predators, too, may encourage diversity.[38]

In most cases of threat we cannot assign guilt neatly between culprits. The introduction of exotics may be a matter of taste in household pets or garden design. It may be a matter of technology, as with crop and livestock preferences in farming. Population growth in Europe was a factor in bringing immigrants with their exotics into the Americas, Australasia and many islands.

The IUCN's *Red Data Book* of endangered plants lists no less than twenty-one separate types of threat. They include clearance of forests, drainage of wetlands for farming, mining and quarrying, urban and industrial development, dams, roads, tourist developments. Also overgrazing, changes in farming methods, logging and firewood cutting, introduction of exotic species, disturbance by trampling or cars, fire, eradication of poisonous species, and collection for flower vases, greenhouses or gardens.[39]

To cover animals and aquatic species they might have added hunting for trade or subsistence, overfishing, water pollution, acid rain and what could become the biggest threat of all – global climate change (see Chapter 15).

What we are dealing with, in other words, is not a simple process with just one or two causes. It is a massive onslaught, firing with all barrels from many directions at once. All our human needs, at every stage of development from hunting through to the affluent society, threaten to drown a great deal of the natural diversity we have inherited. And the more of us there are exerting these needs, the greater is the threat.

Only when we come to value that diversity as one of our human needs – and one that overrides many others – will we reverse the tide.

5

THE PARAGON OF ANIMALS:
Ranomafana forest, Madagascar

Madagascar illustrates, better than any other place on earth, the creative processes of the past – and the destructive forces of the present.

The island confounds all expectations. Come looking for Africa and you will be surprised by Asiatic faces, teams of men digging and turning the soil of paddy fields, rows of women transplanting rice seedlings. But this is not Asia either. The soils are typically African: rust–red laterite on the hills, heavy grey clays in valley bottoms. The faces are darker, and the style of cattle-rearing is African. Madagascar is Africa and Asia fused. Or rather it is neither, but only itself.[1]

Biologically the island is equally unique. Two hundred million years ago it joined Africa to India in the supercontinent of Gondwanaland. Then the underlying plates began to rift apart. India moved northwards and collided with Asia, raising the Himalayas. Madagascar slowly rafted away from continental Africa.[2]

The wildlife is a living record of continental drift. It has elements of Africa and India. Its long separation created new species, and preserved primitive characteristics that died out elsewhere. The lemurs – along with the tarsiers of South East Asia – are the only living relatives of our earliest prosimian ancestors. The hedgehog-like tenrecs are close to the first insectivorous mammals that scampered between the lumbering feet of dinosaurs.

Many animals you might expect to find are absent. There are none of Africa's antelopes, deer, horses, elephants, lions, leopards, or cheetahs.

Yet some families are there in unexpected profusion. The island hosts all the world's lemurs (except for two species it shares with the neighbouring Comoros islands). It has two-thirds of the world's chameleons. Of the bulbous bottle-shaped baobab, which looks like a tree buried head first, roots in the air, Madagascar has eight species, against only one for the whole of continental Africa. It has more species of orchids and of butterflies than the rest of Africa put together.

Because Madagascar has been divided from the mainland for as long as 180 million years, many species are found nowhere else in the world. Twenty-six of the twenty-eight primates are unique to the island. So are thirty-one of the thirty-two insectivores, all seven carnivores and all ten rodents, over 8000 of the 10,000 flowering plants, 142 of the 144 amphibians, and 231 of the 257 reptiles. Many families of animals, finding themselves without competitors, have radiated into a multiplicity of niches. The vanga shrikes range from a hornbilled variety like a small toucan, through shortbilled fruit eaters, to the sicklebill with its long curved beak for extracting insects from tree holes.

The lemurs are equally diverse. Since there are no woodpeckers in Madagascar, the aye-aye fills the niche. The three species of bamboo lemur carefully subdivide their food resource. The grey gentle lemur eats mainly leaves. The great gentle lemur, *Hapalemur griseus*, is the only one that can eat woody stems, helped by symbiotic bacteria living in its gut. The growing bamboo tips that the golden bamboo lemur eats carry enough cyanide to poison a man. But the animal has a special type of blood haemoglobin that can absorb the poison without damage.

This immense biological wealth faces dangers of unequalled severity.

The past serves a clear warning. Within the past two thousand years giant lemurs roamed the mosaic of forest and wooded savannah that once cloaked the plateaux. There were dwarf hippos and a giant tortoise with a shell well over a metre long. And huge elephant birds like the towering *Aepyornis*, chest like a wine barrel, thighs like a horse's, egg big as a football: probable source, through sailor's tales, of the legendary Roc that carried Sinbad off in its talons.

The plateaux are now bare and increasingly barren. The bones are all that remains of the creatures that once lived there. A row of twelve sad skulls in a glass cabinet in Tsimbazaza zoo commemorates the extinct lemurs. The skeletons of *Aepyornis* and dwarf hippo, and the shell of giant tortoise, stand beside them. No major climatic changes occurred that could explain their disappearance. But, some time during the first centuries of our era, longboats sailed over from South East Asia, by way of southern India and East Africa, bringing the first humans to the island.

Within a thousand years of their coming, no land vertebrate heavier than 12 kilogrammes survived. Their habitat, the plateau forest and savannah, was destroyed by fire, turned into pasture for the longhorned, humpbacked Zebu cattle which the settlers brought over, and rice paddies in the valley bottoms. The survivors were hunted to extinction for their meat.[3]

A new wave of extinctions may be imminent. The main threat is clearance of the rainforests to provide farmland for growing populations.

The forests have been reduced to a narrowing strip clothing the eastern slopes of the escarpment that forms a north–south backbone to the island. Of the original forest cover of 11·2 million hectares, only 7·6 million hectares remained in 1950. Today this has been halved to 3·8 million ha.[4]

Every year 111,000 hectares more are cleared. At this rate all of Madagascar's rainforests will vanish entirely within thirty-five years.

Bamboo lemurs

The lemurs, perhaps the most elegant group of primates in the world, are also among the most threatened. Three species – the golden and great gentle bamboo lemurs, and the aye-aye – have only a few hundred specimens each.

Primatologist Pat Wright, gumbooted against hungry leeches that line the paths, pointed out signs of the lemurs' passage through Ranomafana forest, on the eastern escarpment. The ghostly nocturnal aye-aye had poked holes in rotten tree trunks, rooting for grubs. Droppings like balls of wood chips, under sawn off bamboo stems, marked the route of the great gentle lemur. The golden bamboo lemur had chewed through the growing bamboo tips, forcing them to sprout a thicket of new shoots.

Over the next few days I began to spot the lemurs themselves and other animals. A sleepy bundle of three tawny woolly lemurs, clinging in the fork of a tree. A family of ruddy redbellied lemurs grooming each other, far up in the highest branches. Black-and-white diademed sifaka, mushroom-hunting among the leaf litter, staring at me with their disturbing bright red eyes. A tiny nervous brown mouse lemur, steeling herself against torchlights to snatch a banana speared on a branch end. The deep red mongoose, galidia, slunk along the trails, waving its black-ringed bush. Phelsuma geckoes, red patches on their turquoise-green bodies, mated on tree trunks.

Despite several nocturnal jaunts, I didn't see an aye-aye. The great gentle lemurs were so timid that, even when my guide played mating calls, they came no closer than a hundred metres.

The golden bamboo lemur *Hapalemur aureus*, was more compliant. After I'd learned to walk silently, stopping to listen every few metres, I noticed him by a barely audible clipping of teeth. He was stripping a bamboo tip, among the upper leaves of a great arching stem. He spotted me, but didn't run. Instead he shinned boldly down the trunk to inspect me from close quarters, allowing me to admire his plump, dark-brown body, beautiful brown face with a lighter ring circling his big round eyes and nose. Then he tumbled off between the stems, racing hand over foot along the bridges of giant bamboo, leading me below into an impenetrable knot of undergrowth. Thin lianas and thorny vines caught round my neck, tangled in my camera strap. I could only exit backwards.

The golden bamboo lemur is so rare that it was not discovered until 1986, almost by accident, by two expeditions searching for the great gentle lemur *Hapalemur simus*. *Simus* was last seen alive in the wild in 1900, and was feared extinct, but sure enough they found her too.

Both species went straight on to the 'endangered' list. Based on the area of habitat left, there were probably no more than 400 hundred of each. And like the panda, their reproductive rate is slow. The monogamous *aureus* couples have only one baby per year.

Farmers had already cut down most of their forest habitat. And what little remained was threatened. Logging licences had just been awarded for the area. Every day 400 men were going in with hand axes. Desperately poor, working for less than $2 a day, they soon felled every last saleable specimen of three prized tall hardwoods. They lugged them out to the road on their shoulders. From there lorries took them to the coast for export. The severed stumps remain, so soft you can poke a stick right through, hosts to inheriting forms of life – polypody ferns, trailing mosses, bracket fungus, colonies of ants and termites.

The primitiveness of the logging techniques was a blessing. The forest can recuperate: seeds and saplings of the logged species are slowly regrowing.

But the threat from local farmers was more serious. They were pushing back the edges of the forest, clearing land to grow crops.

In 1973 Ranomafana forest was 60 kilometres wide. By 1987 it had been cut back to a strip only 7–15 kilometres across.

The burning

All around Ranomafana forest, villages have been pushing inwards, edging back the limits.

Ambodiaviavy, a few dozen mud-walled, straw-roofed huts on a knoll, looks over flat irrigated paddy fields towards a semicircle of hills to the north.

The valley bottoms are cleared for paddy rice. The lower slopes grow hill rice, cassava, bananas and coffee, or sport fallow stands of shrubs and the elegant splayed fans of Traveller's Palm. Only the precipitous top third is forested. Even there, the lemurs' peace is disturbed by village cattle, grazing on the undergrowth. Once a week, on a Sunday, the men go in to hunt birds, freshwater crayfish, and honey.

The forest frontier is retreating. This year's new clearings bite into the fringe. Patches of matchstick logs tumble downhill, waiting for the torch. One belongs to thirty-seven-year-old Zafindraibe, the man on the cover of this book.

In this moist environment there's no danger of forest fires getting out of hand. Cut rainforest doesn't burn easily. Zafindraibe's plot had been drying out for a month. He left it till the afternoon so the sun could evaporate the dew. At two he set off, in his vest of woven straw and two pairs of shorts, the outer ones torn like an old rag. His legs were spattered with grey mud from the paddy field he'd been digging over with his cousin.

I followed him, across the brook where laundry was drying on the grass; up the stream bed; through the fallow where tall ferny longoza towered fifteen feet high; on through groves of overgrown coffee, shaded by tall albizza trees, leafless in the dry season.

Then we came to the new clearing. It took three men six days to fell the trees. They lay collapsed in untidy piles, in full leaf, with their tangle of liana vines, like a huge arrangement of dried flowers.

Zafindraibe's thickset wife, Baolahy, was waiting with their three children, in the shade of a shelter of banana leaves. He took three smouldering sticks from the fire where they'd cooked lunch, and headed for the lower edge of the pyre. He crumpled kindling of dried bracken into a gap. Puffing on the brands till they glowed bright red, he held them against the bracken, and blew again till the fire took. Leaping hastily over the trunks to escape the flames, he started fires in three other places till half the hillside was ablaze. The fumes, blown uphill, choked the trees on the forest margin.

Zafindraibe had burned the eastern half of his plot the previous week. The main trunks were still left, a hurdle race of blackened poles up to two feet thick. The ground was thick with ash. Baoloahy, with eleven-year-old daughter Zafitsara working beside her, stabbed holes in the ground with a digging stick, dropped in three grains of rice to the hole, moved on.

Across the valley, the smoke from four other fires like this one trailed slowly upwards towards the clouds. Every day, in the dry months of the southern spring, the forest burns. Every night lines of flame glow red against the black mountains. The sky is eerily illuminated by orange clouds of smoke, as if dawn is breaking all round the horizon.

The view from the village

It's easy to demonize farmers like Zafindraibe. Easy to take the line of one visiting French primatologist, who told me, without a trace of irony: 'There are too many people, and not enough lemurs.' Or, more charitably, to imagine that all that's needed to stop the destruction is education about how important lemurs are to the world.

The truth is more complicated, the solutions harder to find. The farmers are victims rather than criminals. They fell the forest for survival, not for profit. To understand the process, and to have any hope of controlling it, you have to look at it from the village point of view.

Fifty years ago this whole area was dense forest. Since then population growth has led inexorably to deforestation and loss of species diversity.

Until 1947 the people of Ambodiaviavy used to live several miles to the south. In that year there was an insurrection against colonial rule. French troops began rounding up local people. To avoid arrest the villagers fled into the depths of the forest.

Sixty-four-year-old village president Tonga, in red-check shift, remembers:

We built shelters of leaves in the forest. We hid our rice in holes in the ground in the old village. In the evenings we would go down and get some for the next day. But then we ran out of food and had to give ourselves up. They shut us up in the Hotel Thermale at Ranomafana. We were in there for a month. There were epidemics – dysentery, coughs. Many died. When we came out we found the French had burned our old village. We thought it was unlucky to go on living there. We liked it here so this is where we built our new houses. There was an old fig tree just over there – it's dead now. Ambodiaviavy means 'at the foot of the fig tree'.

In the beginning, there were only eight families here, thirty-two people in all. At first they cultivated only the valley bottom, easily irrigated from the stream running down from the hilltops. There was no shortage of land. Each family took as much as they were capable of working.

Over the next forty-three years, village numbers swelled ten times over, to 320, and the number of families grew to thirty-six. Natural growth was boosted by immigration from the overcrowded plateaux, where there is no spare land at all.

Two processes gathered momentum as the population expanded. Starting in the 1950s, the valley-bottom lands filled up until they could no longer feed the growing numbers or provide land for new couples. People then started to clear the sloping valley sides. They moved gradually uphill until, today, they are two-thirds of the way to the hilltops.

Parallel with this expansion of the area came a decline in the size of each family's paddy holding – also fuelled by population growth. There is a limit to the land that can be used for paddy. It is defined by the lay of the land and the flow of the rivers. In the early days new couples could just open up a new section of paddy. But as soon as the potential paddy land was fully occupied, new couples relied on their parents to provide a share. Nowadays, when children marry, parents have to subdivide the paddy land and give them a plot. As a result, holdings in the irrigated valley bottoms have dwindled. Today only a few of them are big enough to feed a family.

The more children in a family, the smaller their share as adults will be. The village chief lives in a small mud hut, looking out over a valley which he once owned entirely. Since then he has had ten children, and given parcels away to each. Today he has only two hectares left for himself. Though he is the wealthiest man in the village in cattle, his sons are among the poorest. They have only half a hectare of paddy each. They moved from prosperity to pauperdom in a single generation.

Some of the valley-bottom holdings are now so small that they cannot be further subdivided. If they want to have any land at all, the children are then forced to open up virgin forest hillsides.

Zafindraibe and Baolahy are in this position. Their small paddy field feeds the growing family for only four months of the year. The forest is sitting there, unused by humans, free land waiting for the first man to get there with an axe. It slopes at an angle of over 45 degrees, and is hard to work. But for the next year it will grow rice. The second year cassava will take over, an undemanding crop that provides calories in bulk. After that the plot ought to be left fallow for at least six or seven years.

In the first year of fallow, tall grasses, low shrubs and longoza take over. By the second there are sapling trees and Traveller's Palms. Three years on there are thirty-foot-high trees. Organic matter builds up again in the soil. When the plot is cleared and

burned again, the ash provides free fertilizer and soil conditioner. As long as the fallow cycle is followed, erosion is low.
The dense fallow cover above and below each cultivated plot
slows down run-off and collects the small amount of soil lost.

Population growth is disrupting the fallow cycle. As land
shortage increases, a growing number of families can no longer
afford to leave the hillsides fallow long enough to restore their
fertility. They return more and more often. Each year it is
cultivated, the hillside plot loses more topsoil, organic matter,
nutrients.

I came upon Marie Rasoanirina burning the stumps of last
year's pineapple plants on a quarter-hectare field. This year she
would plant beans and cassava. Then she would leave it fallow
for two years, instead of the six it needs. She can't afford to
leave it for longer, because two hillside fields is all the land she
has. Her husband is a landless immigrant from the high
plateaux, even more overcrowded than here. He came in when
the logging operations started. Marie's own parents haven't
enough irrigated land to survive, let alone to give her. They are
selling it off, in small parcels, to get money to live on.

Eventually the fallow period is suppressed altogether. Vola
has a tiny plot of paddy, about one-tenth of a hectare, and half
a hectare of hillside land. She was planting rice that year, to be
followed by cassava. After that, instead of fallow, she would
plant bananas and keep them for three years. Then rice again.
Her husband died nine months earlier, when the logging lorry
he worked on crashed. With four children to feed, she couldn't
afford to leave the land fallow for even one year.

Ranomafana forest has been saved by a *deus ex machina*. The
World Wide Fund for Nature began pressurizing for the area to
be made a national park, to preserve the last bamboo lemurs.
The decree was signed late in 1990.

But not everyone lived happily ever after. The pressure
shifted from the forest and animals on to the people. The
farmers of Ambodiaviavy are no longer allowed to clear new
plots when they are short of land. They are banned from
grazing cattle, cutting timber for building, hunting, and most of
their other traditional uses of the forest.

A massive $4·3 million development project has been
mounted to protect the park, and help the surrounding villages

in an attempt to compensate them for their losses. But very few of the funds are earmarked directly for villagers. The overwhelming bulk goes on salaries, fares and facilities for American consultants and students. A small amount goes on local salaries for park guards and guides. Western eco-tourists already stamp the forest trails, while villagers are kept out.

But the pressure to clear more land continues. The first encroachments have already occurred. Ten farmers made a mass invasion inside the forest to try to establish land rights. Ten park guards and six gendarmes swooped and arrested them. But most villagers suffer in silence. With the forest frontier closed, the landless young will move to the cities.

The locals are understandably disgruntled and apprehensive. They can't help feeling that the rich world cares far more about furry animals than about poor suffering humans. Recently the mother and child of the most studied golden bamboo lemur family disappeared without trace, probably eaten by a fossa. There was open grieving in the primatologists' research camp, and puzzlement among the Malagasy guides.

'Why do Americans weep when a lemur dies?' one of them asked me. 'Why don't they weep when a Malagasy child dies?'

There was no real choice: the lemurs and their forest had to be protected. The farmers could have carried on clearing for another two decades at most. At the end of that time the forests would have gone from the watersheds. Torrential rains would flood their fields more often. Erosion would increase. Streams would run for only a few months each year. And the golden bamboo lemur would join its extinct cousins in the zoo museum.

But the cost of preserving the world's wildlife heritage should never fall on the shoulders of the world's poor. The rich North has an obligation to make sure that the world's gain is not their loss. We cannot save the animals without providing fully for the humans around them. We must learn to weep for the people as well as for the lemurs.

Pointing the finger

The ecological tragedy in the making here was cut short. But 999 times out of a thousand, it is played to its destructive conclusion, across Madagascar and throughout the tropical rainforests. The forest is cleared to satisfy the demands of survival. The wildlife retreats or dies out. But the respite is only temporary. Eventually the limits of the forest are reached and the reckoning has to come.

The adaptation that Ester Boserup theorizes about (see p.31) has not happened in Madagascar's rainforests. The land is being mined, without fallow to recover, without fertilizers, without conservation measures. Erosion carries off more and more topsoil as the vegetation cover weakens. The forest no longer acts as buffer, storing rainy-season water, letting it out gradually through the year. Instead the rains wash off as they fall. The dry season stream beds are parched, the wells dry up. Yields decline, and farmers can no longer wrest a subsistence living from the land. The forest and the wildlife are lost, for perhaps twenty or thirty years of miserable survival.

Who or what is to blame for this tragedy?

Consumption levels are minimal. The villagers eat barely enough to stay alive, certainly not enough to keep healthy. Nor is Western consumption a factor. Logging here has been gentle. It may have temporarily reduced diversity and thinned the forest a little. But it has not caused deforestation.

Population growth is the clearest culprit for deforestation and species loss, growing land shortage, shrinking holdings, and impending land degradation. The equation seems simple: the more people there are, the more forest they have to clear to feed themselves.

But technology is also a factor. If the peasants of Ambodiaviavy had used fertilizer or improved seed on their irrigated land, they could have increased food production in line with population growth without clearing any more forest. This would also have alleviated the land shortage: a smaller parcel would have been capable of providing a livelihood for a family. When technology stagnates, the only way of increasing food production is to expand the area under cultivation.

So why didn't they intensify?

If we accept the full Boserup thesis the answer is simple. Intensification involves more work in ploughing, manuring, weeding. It will be undertaken only when it is unavoidable: when there is no more new land to clear.

In this case the situation is slightly complicated. The people of Ambodiaviavy have already intensified on their paddy land, at least in terms of labour input. They have not yet begun to add fertilizer. They do have cattle, but graze them on the free resource of the forest, so the dung is wasted from an agricultural point of view. But on the non-irrigated hillside land they have not intensified at all. And as long as there is new hill forest to open up, they would have no interest in intensifying production on old hill land. Why pay for fertilizer, when the fallow period makes soil nitrogen for free, and the ashes of the burnt fallow trees provide phosphates and potash?

However, the Boserup thesis does not explain everything here. Zafindraibe knows – having visited a development project down the valley – that with only 60 kilos of fertilizer, he could get enough rice to last his family all the year – a gain worth many times more than the $20 cost of the investment.

So why didn't he buy fertilizer, I asked?

Because there was none for sale locally.

Would he buy any if it was available?

He couldn't afford to, because he had no cash.

So why doesn't he borrow and repay from the profits?

'No-one in the village will lend,' he explains. 'If there's not enough rain, or if insects attack the crop, they're afraid they won't get their money back.'

Why not borrow from a bank then?

'If a flood swept away my crop, I couldn't repay the money. I'd be put in prison, or they'd take my land away.'

Then I noticed that he'd let his coffee bushes grow into tall straggly trees. Pruning back would give much higher yields.

He knew he should prune. But the bush wouldn't produce the year it was pruned, and he couldn't afford to pass up the income.

So poverty – lack of cash to buy fertilizer, lack of savings to allow pruning – helped to keep technology stagnant. The hazards of climate and pests made borrowing and lending risky.

The international economic system played its role too. Over the past fourteen years the price received for coffee had fallen by two-thirds. To reap the same income, they would have to triple the area under coffee, or expand the area under rice to make up the loss of cash income. Lowered commodity prices were also pushing towards deforestation and degradation.

Other aspects of the international economy conspire to keep Madagascar's farmers at a low technological level. Unprocessed agricultural products, with their volatile prices, still account for more than four-fifths of its exports. Like so many developing countries, Madagascar built up huge debts – rising from $956 million in 1980 to $3317 million in 1988. Debt service, which used up less than 4 per cent of export earnings back in 1970, swallowed almost ten times that proportion in 1988. The country had to rein back imports – including chemical fertilizer. It also had to slash government spending savagely, from 23 per cent of gross national product in 1965, to only 12 per cent in 1988. Agricultural extension was among the casualties.[5]

Market imperfections were also to blame in holding agriculture back. From the early 1970s the state took control of marketing and prices of food and many agricultural commodities. Most of the profits of the coffee boom of 1977 made their way into government coffers rather than the farmers' pockets. In 1981 and 1982, more was spent on subsidizing rice for urban consumers than the total budget of the Ministry of Agriculture.[6]

Population growth is certainly of crucial importance. But it is not a purely independent factor. It is influenced by other aspects of the situation. Population growth is high wherever health and education are poor and family-planning supplies are inadequate.

Four out of five children in Ambodiaviavy are not in school. Villagers complain that the teacher hits children and is absent two or three days a week. Teachers' salaries start at £5 a week, a day labourer's wage, so good staff are hard to attract or keep. Parents also keep children at home to look after animals or younger children, or to help out with work. In the hungry months before the harvest, children often have to look after the home while parents are away trying to earn money to buy food.

High infant mortality keeps the birth rate high. Parents have extra children to insure against probable losses. Statistics are unreliable, but most experts believe that infant mortality in Madagascar has actually risen since 1975, to roughly 125 per 1000 live births in 1990, because of declining incomes, rising malnutrition and cuts in health services. Infant mortality in the villages is certainly much higher still, probably at least 200. It is not uncommon to meet women who have lost half of all their live-born children.

The health of Ambodiaviavy's people is among the worst I have seen in seventeen years of travel throughout the Third World. Half the children were infected with malaria, though only one in six had had fever in the past fortnight. The children delouse one another in lines or circles of four or five. The village is riddled with fleas. It took me three days to get rid of the ones I caught. One in six children has scabies lesions on their hands. One in three has lesions from the jigger flea. The female eats her way into flesh, covers herself with a cyst, and converts herself into a living brood chamber, bloated with her swelling eggs. The hatched larvae eat their way out. Locals pick the cysts out with a pin – but the sore often gets infected.[7]

There are internal parasites. Over 90 per cent of the children have an average of six roundworms, as big as a medium-sized garden worm, living in their stomach. One child had a hundred, with a combined weight of 2 kilos. Half have whipworm as well, and a third have hookworm. These parasites consume much of the limited food that the child eats. This contributes heavily to child malnutrition. So does diarrhoea – one person in three has an attack in any given fortnight. Almost six out of ten children are malnourished – one in ten severely so.

There are the bleak cases like forty-eight-year-old-Fambelo, who hobbles around on a stick, no longer able to dig his fields, with swollen, aching throat and back pains. He has been seriously ill for a year but hasn't seen a doctor, because he's afraid of the cost in drugs and hospital charges. Blind, landless Miray can afford no treatment or help, but supports his five children working in others' fields, feeling his way.

The health situation is getting worse because of government cuts. The oldest resident, Zafindraibe's seventy-two-year-old father Ravanomirina, remembers when medicines were cheap, doctors were plentiful, and insecticides kept down the fleas and lice. 'Nowadays,' he complains, 'people are afraid of going to the doctor because they have no money to buy medicines. Even when someone cuts their leg open, they don't go to the hospital in Ranomafana. They just pour hot water on the wound and leave it.'

Despite all these obstacles, there would still be some limited demand for family planning, especially among poor women like Vola who can't afford to feed their existing children adequately. But the nearest supply is at Fianarantsoa, four hours' journey away by bone-shaking bush taxi. The return fare amounts to four days' wages. Family planning services are almost non-existent in Madagascar – in 1990 less than 2 per cent of fertile women were using modern methods.

So where can we lay the blame?

It would be quite easy to come away from Ambodiaviavy with a Malthusian view that population growth was the sole cause of deforestation and loss of species. But it would be an oversimplification. For if technology gradually improved yields in line with population growth, then there would be no need to clear more forest to feed more people.

It would be equally easy to come away convinced that technology lag, due to individual and national poverty, was to blame. These in turn are aggravated by declining commodity prices and an intolerable debt burden. Or one could blame the lag, the debt and the poverty on misguided government policy, which controlled markets and prices paid to farmers, and made it uneconomical to intensify.

But these, too, would be oversimplications. For if population were not growing, then there would be no need to clear new forest, even if no fertilizers or high-yielding varieties were used.

Deforestation and the resulting species loss happen because population growth is outstripping change in technology. The speed of both relative to each other is what counts. It is meaningless sophistry to claim that one is more fundamental than the other, or to call the one a 'root' cause and relegate the other to a mere 'exacerbating factor'.

Population growth and technology are of equal importance in explaining deforestation and species loss – and there are many other factors that explain each of these. We have to understand the role of all these factors together, and tackle as many of them as we can, before we can hope to see significant reductions in either environmental damage or population growth rates.

6

THE GRINDING OF THE AX:
deforestation

There is not enough brushwood –
It is the fault of the people of Wei.
They have wasted the land with fire.

Song of Hu-tzu, On the flooding of the
Yellow River, c 109 BC[1]

Zafindraibe's bonfire on the Eastern escarpment of Madagascar is part of a global conflagration. On night-satellite photographs burning forests show up brighter than city lights at certain seasons of the year, from South East Asia, through the Sahel, to Central America and the Amazon.

The sheer scale of deliberate deforestation is recent, but the phenomenon itself is as old as agriculture. Cereal crops are grasses, and prefer open terrain. Pastureland has to be kept open, and free from invading shrubs. Fire and the axe are inseparable from pastoralism and agriculture.

The Mesopotamian epic of Gilgamesh, from the third millennium BC, is in part a myth of deforestation. The hero Gilgamesh represents the agricultural state, his hairy mountain friend Enkidu pastoralism. Together they gaze in wonder as they approach the vast cedar forest of Lebanon. It stretches for ten thousand double hours of travel, its trees are six dozen cubits high. But the pair have not come as nature lovers: forest is anathema to both of them. They have come to kill the forest's fierce guardian, Humbaba. They smoke him out with a forest fire, take him prisoner, then kill him despite his pleas for mercy. Then Gilgamesh fells the trees and Enkidu digs up their roots as far as the banks of the Euphrates.[2]

By classical times much of the Mediterranean was extensively deforested. Most forests were cleared for farming. What remained was plundered for fuel and timber for building houses, ships, military camps. As early as the fifth century BC Attica had to import all its wood for shipbuilding. Plato complained in the early fourth century BC that hillsides which once grew huge trees had been turned into open heath fit only for bees (see p. 115).[3]

Northern and western Europe's forests survived longer, protected by their heavy clay soils. These were impossible to work with the Mediterranean scratch plough, which merely scrapes the surface. A heavy plough was needed, which cut the sod and turned the soil over. Once this spread, from the early centuries of the present era onwards, the region was progressively deforested. Today only one small reserve of a few square miles, on the borders of Russia and Poland, still represents the flora and fauna of the original forest.[4]

Early settlers in North America regarded the forest as a dark, hostile wilderness harbouring dangerous animals and Indians. By 1860 perhaps 60 million hectares had been cleared for farming, plus another 5 million by logging, mining and urban development. At the end of the nineteenth century there were virtually no new areas of virgin forest left unexploited. As late as 1920 logging was proceeding at nine times the rate of regrowth. It was not until 1963 that the annual regrowth exceeded the amount cut.[5]

Today the focus of deforestation has shifted to the South. The standard TV nature programme from the tropics follows an obligatory sequence. It begins with cute animals and curious plants, and ends with buzzing chain-saws. These emotive images heavily colour our perceptions. The rainforests, we are told, will all be gone within a decade or two.

The real situation is not quite as bad as this – but it is bad enough.

The global loss of forest since the beginning of agriculture is significant. But it is a good deal lower than most people might expect. NASA scientist Elaine Matthews attempted to estimate the world's original forest cover, based on the areas that were climatically suitable. Comparing this with the situation around 1970, she concluded that the original cover had declined by 15 per cent. Another study by R.A. Houghton and colleagues, taking different start and finish dates, suggested that 15 per cent of the forest cover of 1850 had been lost by 1980.[6]

Experts disagree on how much forest remains. Recent estimates range from a low of 37·6 million square kilometres right up to 52·4 million km^2. Land-use figures from the Food and Agriculture Organization suggest that there were some 40·5 million square kilometres of forest and woodland in the world in 1988. This was almost one-third of the entire global land area. It dwarfed the 14·75 million square kilometres of farmland, and comfortably outstripped even the 32 million km^2 of pasture.[7]

The developed countries have a greater proportion of their land area covered with forests − 34 per cent. In developing countries the proportion is only 29 per cent. The bulk of the Third World's forests are concentrated in a handful of countries. Brazil alone had 5·6 million square kilometres in 1988 − more than Zaire, China, Indonesia, Peru and India, the next five countries, put together.[8]

The lush tropical rainforest is even more clustered. Just three countries − Brazil, Zaire and Indonesia − have more than half the world total. Latin America has 56 per cent. Nine-tenths of that is in the Amazon basin.[9]

There is little agreement over the rate of deforestation. Assessments fluctuate wildly at country level according to definitions of forest and non-forest, methods of measurement, and whether logged-over areas are classed as deforested. Recent figures for Brazil vary between 17,000 and 80,000 square kilometres per year. For India during the 1980s they stretch from a low of 475 square kilometres a year right up to 15,000 km^2.[10]

At global level the divergence between estimates is less dramatic, but still pronounced. Figures for the area of tropical forests lost each year range from 133,000 square kilometres up to 245,000 kilometres.[11]

The FAO's land-use figures give some idea of trends over time, and the results are unexpected. Contrary to the prevalent image, at a global level deforestation has slowed considerably, from 112,900 square kilometres per year in 1973–8, to only 71,300 square kilometres a year between 1983 and 1988. Over this period the rate of loss of world forest dropped from 0·27 per cent a year to only 0·18 per cent.[12]

This slowdown is due entirely to trends in developed countries. In 1973–8 they were still losing some 27,000 square kilometres of forest per year, but ten years later their forests were expanding by an annual 36,000 square kilometres.

In developing countries almost all studies suggest that the rate of deforestation has been accelerating. In some countries the losses are dramatic. In Madagascar only 34 per cent of the original forest cover is left. In the Côte d'Ivoire and the Philippines four-fifths of the original forest has been cleared. Ethiopia's forest cover dwindled from 40 per cent of her land area in 1940 to only 4 per cent fifty years later.[13]

The most complete survey to date was published by the Food and Agriculture Organization in 1982. This suggested that tropical forests of all types were being cleared at an annual rate of 113,000 square kilometres around 1980 – losing just under 0·6 per cent of their area each year.[14]

The highest rates of loss were found in countries where populations were growing rapidly, but where forest areas were neither very large nor very small. Central America, Costa Rica, Haiti and El Salvador were losing more than 3 per cent of their forests every year, while forests in Nicaragua and Honduras were shrinking by 2·5 per cent a year. The other region of extremely high pressure was West Africa, where Côte d'Ivoire, Benin and Nigeria were deforesting at the rate of 3·7 per cent a year or more. In Asia only Nepal was losing forests at such headlong rates.[15]

Lower rates applied, despite rapid population growth, in areas where most of the original forest cover had already been cleared for agriculture – in South Asia, China, Rwanda, Burundi and Kenya.

Even lower rates were found in countries with vast areas of forest. Here large amounts of clearing made only a small impression. Brazil, for example, accounted for a third of Latin American deforestation – yet was losing only 0·27 per cent of her forests each year. Zaire had the third highest annual loss in Africa, in terms of area – yet this amounted to only 0·15 per cent of her gigantic forest area.

The lowest rates of all were found in countries with small populations and very large forest areas – such as Congo and Gabon, the three Guyanas and Papua New Guinea. These countries were losing less than 0·1 per cent of their forests each year.

In 1990 the FAO began a new assessment of tropical deforestation. Though the methods and coverage were not strictly comparable, preliminary results suggested that tropical deforestation had accelerated during the 1980s, to around 170,000 square kilometres – 1 per cent of the forest area – each year. This amounts to two-thirds of the area of the United Kingdom. Every ten weeks an area the size of the Netherlands. A Barbados every day.[16]

What's to blame?

The familiar dispute rages over the causes of deforestation. Population growth, shout the neo-Malthusians. Logging, ranching, unequal landholding, everything but population, shout the critics. Let's look at the factors one by one.

Logging for timber is most frequently in the dock. Rainforest hardwoods, which evolved to withstand perpetual damp, aggressive insects, and ubiquitous fungi, are long lasting. They are in great demand for doors, windows, heavy furniture.

By 1985, some 1·9 million square kilometres of the total area of 11·25 million square kilometres of closed tropical rainforest had been logged over. The proportion varied widely between continents. Of the vast reaches of the Amazon, less than 9 per cent were logged. In Africa the share rose to 20 per cent. The rainforests of South East Asia had been most heavily plundered for timber. Here 30 per cent had been logged. Many areas, including the Philippines and peninsular Malaysia, had been exhausted of their valuable species by the late 1980s.[18]

Logging need not always degrade the forest environment. Long intervals between cuts can allow forests to regenerate naturally. Felling can be done in the direction that causes least damage. Well-planned extraction methods and roads can minimize damage to forest soils. Human intervention would then have little more impact that natural events such as fires and hurricanes which periodically clear swathes of forest, and are followed by regeneration.

But most logging operations are not designed to maintain the

forest's original diversity. In the tropics they are usually
designed to extract maximum profit in minimum time.
Probably less than 25 per cent of the world's forests were being
managed in the early 1980s. Three-quarters of this total were in
Russia.

In tropical areas, only 20 per cent of logged forests were
actively managed in 1980. Most of these were in Asia, where
management usually meant no more than the existence of an
official concession. In Africa only 4 per cent of forests were
being managed, mainly in Ghana and Uganda, and in Latin
America a quarter of 1 per cent. British forestry expert Duncan
Poore recently estimated that no more than 10,000 km^2 of
productive tropical forests out of a total of 8.3 million km^2 were
being managed on a sustainable basis in 1985.[19]

Careless logging with heavy machinery leaves a trail of
environmental devastation. In an average rainforest, only ten or
fifteen species interest the logger. Yet there may be only a
handful of individuals of these valuable species in one hectare,
growing cheek to cheek, interlinked by climbers and creepers,
with hundreds of other species of no commercial interest. When
loggers take out the biggest commercial stems, their clamorous
fall topples or scars many of their companions. In west
Malaysia loggers take only three per cent of trees, but severely
damage more than half. In another Malaysian forest extracting
10 per cent of the trees damaged another 55 per cent – leaving
only 35 per cent undamaged.[20]

Access roads, bulldozers and dragging logs bare and compact
the soil. Taking out only eleven trees per hectare in East
Kalimantan left 30 per cent of the ground bare of vegetation.
Rain runs off instead of filtering in, and there is heavy erosion
where before there was none. The roots of logged trees decay,
and landslides may follow on years later.

The soil that is washed away fouls streams and rivers, fills up
dams and canals downstream. Logging increased the silt load of
streams thirty-three-fold in one Kalimantan study.[21]

All this has a harsh impact on wildlife. Food no longer falls
into streams from overhanging trees. Fish accustomed to clear,
well-oxygenated water die as the water muddies. Monkeys
starved of fruit often suffer high mortality, leaving gaps in
future breeding populations. The Amazonian monkey *Chiropotes*

satanas was eradicated by the removal of only two or three trees per hectare, since this damaged up to 60 per cent of other trees.[22]

Logging has long-term effects on wildlife diversity. It removes the tallest stems, and reduces the height of the canopy. Living space is compressed into fewer layers, competition increases, and there may be room for fewer species.

The make-up of the forest changes. Rainforest seedlings have adapted to growing in shade, shooting up when an old tree collapses and leaves a gap in the canopy. The bigger clearings left by logging are often invaded by species used to growing in lighter conditions. Where the soil has been scalped, tree seeds may be missing, or unable to break through the compacted surface. Forests that have been repeatedly logged can degrade to scattered patches of original forest, peppered with dense clumps of bamboo spreading from rhizomes, and creeper-covered clearings where topsoil has been scraped away.[23]

Forest cowboys

For all its damage, logging, of itself, is hardly ever a cause of complete deforestation, in the strict sense of converting forest to non-forest. This can happen in rare cases where the forest is clear cut over large areas, too far from remaining stands for forest seeds and animals to reinvade.

Given time to recover, a logged forest will become forest again. Species diversity will be lost, in some cases forever. But the regrown forest can still serve most other environmental functions, protecting soils and watersheds, regulating river flow, slowing global warming by fixing carbon.

Only conversion to other uses — farming, pasture, towns, roads — can cause permanent deforestation.

So let us turn to our second suspect: ranching. Another popular image has the cleared rainforests filled with the whoops of cowboys and the lowing of cattle destined for mincing into hamburgers for the West. To get this into perspective we have to look at the land-use figures again. Over the whole of the Third World, pastureland expanded by a total of 103,500 square kilometres between 1973 and 1988, equivalent to only 7 per cent of forest loss over this period. It does not seem, then, that ranching is a major player on the world scale.[24]

But the impact varies dramatically between the continents. In the Near East and Far East, there was practically no change in the area of pasture over the fifteen years to 1988: no hamburger connection there. In most of Africa rainforest ranching is difficult and costly because of the tsetse fly, carrier of the fatal livestock disease trypanosomiasis. Indeed, Africa's farmland has expanded at the expense of pasture as well as forest. Pastureland in Africa shrank by almost 6 million hectares between 1973 and 1988.

In Latin America, by contrast, ranching is the biggest single factor in deforestation. The region lost 750,000 square kilometres of forest over this same period. The pasture area expanded by 330,000 square kilometres. This easily outstripped the 280,000 square kilometre increase in arable area.

If we assume that one out of every three hectares of new pasture came from conversion from former grassland, then ranching can be blamed for around 200,000 square kilometres of deforestation in the decade and a half up to 1988 – or 27 per cent of total deforestation in Latin America.

In the Amazon basin, cowboys drive herds of humpbacked cattle down rust-red mud roads, between devastated fields where the skeletal remains of giant Brazil nut trees tower. The sight is deeply disorienting. And that feeling is not misplaced, for no habitat is less suited than tropical rainforest to extensive livestock-rearing. The appearance of vibrant fertility is a deceptive illusion. The forest lives on itself, not on its soils. It draws nourishment from layers of leafmould, through surface roots. Most underlying soils are poor, leached of nutrients by millions of years of heavy rainfall.

The cleared forest offers poor sustenance for crops, livestock, or poor peasants. Crop yields in cleared Amazonian rainforest quickly fall to the point where small farmers, using no fertilizers, can no longer get a living from the land. They sell out to large landowners, who graze cattle. There are fewer jobs – one cowboy per 300 hectares, against one farm worker for three hectares. And very soon there are fewer cattle too. The productivity of grass tails off. At first one animal can be maintained on each hectare. After five years of grazing the stocking rate falls to one animal for every four hectares. Ranchers who want to keep up their herd numbers are forced to expand relentlessly. The notorious land disputes and murders of the Amazon happen when neighbouring smallholders or rubber tappers refuse to sell. The falling fertility of the soil fuels social chaos. Cutting down trees for cattle leads to cutting down people.[25]

The tragedy is that so much deforestation in Brazil was made possible only by state subsidies. By 1985, 631 big ranches had received tax credits that could be set against other liabilities. This attracted many urban businessmen. Ranchers also got cheap government loans at 85–95 per cent discount. After allowing for inflation, the interest rates were negative – the Brazilian government was paying ranchers to borrow money. A financial analysis found that without government subsidy many ranches would have made heavy losses. The Brazilian taxpayer was shelling out billions to make a few rich people richer, at the expense of destroying the rainforest and the livelihoods it sustained.

Population, policies and technologies

The argument usually runs that it is greedy Northern consumers and grasping multinationals who are responsible for logging and ranching. This case has been grossly overstated. The truth is that even logging and ranching are connected with population growth in developing countries.

Meat is eaten in homes in Latin America, as well as in Northern fast-food outlets. The infamous hamburger connection, indeed, is a mere trickle compared to home consumption. In 1989, for example, Latin America had a herd of 313·7 million cattle, yet her net exports were only 0·8 million head live plus 100,170 tonnes of meat. Brazil had 137 million head of cattle, but exported only 1931 head, against imports of 122,000. Her net exports of meat were only 120,000 tonnes.[26]

Wherever meat is consumed, it is consumed by people. The total area used for pasture is a simple function of population, times meat consumption per head, divided by the yield of meat per hectare. Other things being equal, an increase in human population, and or meat consumption, necessarily leads to an increase in the area needed for pasture or fodder.

However, rainforest ranching is mainly determined by other considerations. It is ecologically unsustainable and economically irrational. Only government subsidies make it profitable on any scale. These – not Latin or Northern meat consumption – are mainly to blame for rainforest ranching.

With timber, too, the 'hardwood frontdoor connection' has been exaggerated. It is not just thoughtless Northern yuppies with mahogany windowsills, or reckless Japanese builders with wooden scaffolding, who need wood from tropical forests. Timber and other wood products are a basic consumption need that is in increasing demand in developing countries too, as populations and income levels grow. The great bulk of Third World timber production now goes to meet the timber needs of the Third World. In 1988 home use of sawnwood in developing countries exceeded exports by eight to one, and in the case of industrial roundwood by almost eleven to one. And as populations and incomes in developing countries grow, the export share is dwindling further. Net exports of industrial roundwood from developing countries amounted to 11 per cent of production in 1977, but only 1 per cent eleven years later. If these trends persist, the Third World will become a net importer of timber during the 1990s.[27]

As with pasture, the area that needs to be logged is determined by the number of people, times wood consumption per head, divided by the yield of wood per hectare. Most tropical forests are not planted, but simply harvested: nature determines the rate of tree growth. The yield of the forest depends on sawmill technology – how much usable wood can be got from each tree – and even more on consumer tastes. If only four species are in demand, the yield will be very much lower than if consumers will accept eight species. The area that has to be logged will be correspondingly higher.

The amount of damage done by logging to a given area of forest is a direct function of technology. The greatest damage to bio-diversity comes with clear felling and replanting with a few commercial species; or with plantation crops like oil palm, where birds and mammals are poisoned, shot and trapped to stop them damaging the trees.

The direction in which trees are felled makes a huge difference. One experiment in Sarawak, for example, felled trees in the direction that caused least harm. Damage to other trees was cut by half, and the area of forest that became open space fell from 40 per cent to 17 per cent. The way in which the cut trees are extracted is important too. If they are skidded along the ground by tractor, soil disturbance can be three times higher than if skylines are used.[28]

Nature's capacity to regenerate is immense if it is given the chance – and here, too, technology has a big impact. If the areas logged are not too large, and refuges are left all around them from which plants and animals can recolonize, then damage to bio-diversity can be minimized. The rotation period is important: if loggers return to a logged patch too soon, many species may not have matured.

Technology, in turn, is heavily influenced by government policies on forestry. In most developing countries forests are state-owned, and governments grant logging concessions. In theory the state has the power to regulate and tax logging so as to minimize damage. In practice this hardly ever happens.

Royalties and licence fees are usually nowhere near high enough to encourage wise use of the resource. A World Resources Institute study of forest policies found that logging in the tropics produced high levels of rents or excess profits, ranging from $7 to $98 per cubic metre. Government levies captured only 17–38 per cent of these windfall gains: the rest went to concession holders. Between 1979 and 1982 Indonesia lost almost $3 billion of the potential rents to private loggers.[29]

In many cases concessions are awarded to cronies and relatives, in blatant cases to forestry officials or politicians themselves. Short concessions are the rule. Rarely does a concession last as long as the regrowth period of the trees. Loggers have no interest in ensuring that healthy regrowth occurs. This, plus the windfall profits to be had, encourage loggers to view the forest as pirates would treat a captured galleon: a prize to be swiftly looted and abandoned. The natural capital built up over centuries – and in the case of bio-diversity over millions of years – is an Ali Baba's cave to be stripped bare overnight.

The pillaging of the rainforests is not the result of ignorance or incompetence. In many tropical countries, the rainforests serve as political pork barrel. They have been plundered for quick gains by politicians and their protégés. In most of Latin America, they are sacrificed to the landless as an alternative to land reform. In Asia and Africa they offer an alternative to creating urban jobs or investing in rural areas.

And there are few political costs. The rainforests are populated, if at all, by hunter-gatherers. They alone possess the complex knowledge and culture needed to understand the forest, to survive in it without reducing its diversity. Yet they live at very low densities and their total numbers are small. And they are usually politically marginalized and illiterate. They have no legal title to their land. Often they don't even speak the official language. No group in the world is easier to expropriate.

States are formed by agricultural peoples and run from urban bases. Forests in such states are used and abused to benefit urban and agricultural interests.

7

ABATEMENTS AND DELAYS:
forest adjustments

The Russian forests are going down under the axe.
Millions of trees are perishing, the homes of wild
animals and birds are being laid waste, the rivers are
dwindling and drying up, wonderful scenery is
disappearing never to return. ... The climate is
ruined, and every day the earth is growing poorer
and more hideous.

Anton Chekhov, *Uncle Vanya*, 1897,
(trs. Constance Garnett)

Our third suspect, population growth, leads to deforestation
mainly through demand for agricultural and urban land.

Farmland in developing countries increased by no less than
487,000 square kilometres between 1973 and 1988, an area well
over half the size of the United States. This was needed to
accommodate the food and agricultural livelihood needs of just
fifteen years of population growth.[1]

Roughly three-fifths of this expansion was accounted for by
shifting cultivators, who farm a plot for a year or two, then
move on and leave it fallow. In 1980 there were an estimated
500 million of them in the tropics, using a total area of 4·1
million square kilometres of forest and woodland.[2]

Shifting cultivation is far more damaging to forest ecology
than logging. Not just a few selected trees, but every tree on a
plot is felled, then burned along with all other ground plants,
epiphytes and climbers and animals that have not got away.
The blaze heats the soil to 150–375°C, and kills roots, rhizomes
and seeds. Organic matter in the topsoil is oxidized.

One study in East Kalimantan compared the impact of
logging versus felling and burning. On the logged area, 154

species were regrowing within six months. On the felled and burned plot there were only 122. Another area that was burned and cultivated twice had only 79 species. All but fifteen of these were invader species that did not belong in primary rainforest. One of the most damaging invaders after repeated cultivation is *Imperata* grass. It spreads by underground rhizomes which can withstand fire, and once established is very difficult to eradicate. *Imperata* occupies as much as 200,000 square kilometres in Indonesia, and is increasing there by up to 1500 square kilometres each year.[3]

Logging's biggest role in deforestation is as facilitator for shifting cultivators. Logging roads provide access into the forest for settlers. Instead of eating away at one outer edge, they can corrode the forest on many fronts from within. Just four logging roads criss-crossing a square block of forest will double the perimeter accessible to settlers. But logging roads do not cause the influx of settlers – they merely facilitate it. Madagascar's forests have been eaten into patches like a ragged quilt – yet there are very few logging roads.

A far greater impact on forests results when trunk roads are driven deep into inaccessible reaches. The Brazilian Amazon was shielded by its vastness and impenetrability. But in the second half of the 1960s, the military government decided to open up the internal frontier, partly to access mineral reserves, partly to offer a safety valve for the landless. The road network increased from only 6000 square kilometres in 1960 to 45,000 km^2 in 1985. Deforestation accelerated with terrifying thrust: as late as 1975, only 0·6 per cent of the Amazon had been cleared. In just five years this quadrupled to 2·5 per cent, and by 1988 had risen to 12 per cent. Much of the deforestation was done by settlers in government-sponsored colonization programmes. But for every official colonist, there were dozens of spontaneous migrants flooding in as sharecroppers, labourers, squatters.[4]

Many other governments have official resettlement programmes, encouraging farmers to clear rainforest. Ethiopia shifted families from the crowded eroded north of the country to forested land in the south. Malaysia has an active settlement programme.

Indonesia has a vast transmigration programme to move

people from overcrowded Java on to the outer islands. The programme has never succeeded in relieving more than a small share of the population pressure on Java. But it has helped to speed deforestation on Sumatra, Sulawesi and Irian Jaya. In colonization programmes governments act as facilitators for a movement that would occur anyway, under the pressure of land hunger from overcrowded areas. Spontaneous settlers outnumber official colonists by anything from five to one up to a thousand to one.[5]

But farming in the rainforest is often no more sustainable than ranching. Most rainforest soils have been leached of their useful nutrients by heavy rainfall, leaving insoluble oxides of iron and aluminium that hamper plant growth. Organic matter breaks down rapidly in the high temperature and humidity. Pests and diseases of crops, livestock and humans proliferate. It can be farmed by mimicking the structure of the forest, using complex multi-storey systems with ground, shrub and tree crops of all sizes. But most new settlers in rainforest simply import the techniques they know from their home areas, and these are usually unsuitable. The soil degrades, the plot is abandoned, and another plot cleared.

On the island of Sylvania

Most people accept intuitively that farmland expansion is due to population growth. But not everyone. Some critics of the population role in deforestation imply that more people need more land to produce more food only if they cannot intensify production on their existing land. Or indeed if they have no land at all. Poverty, insecurity or inadequate prices are what prevent them from intensifying. Unequal landholding is what leaves them landless. Therefore these factors, not population growth, are the real causes of deforestation for farming.

Let us begin to clarify with a thought experiment. Imagine a group of 100 people, driven out of their original home by invaders. After months of perilous sea journeys, they finally spot a beautiful island, entirely covered in rainforest. They beach their longboats and explore. The island is uninhabited. The soil seems fertile. They decide to settle and call it Sylvania.

The elders sit down to work out how much land they need to clear. Now half the land is gently sloping, half steep. The total area is exactly 1000 hectares. And, to the relief of the tribal mathematician, there are just 100 people.

Now over the centuries of their traumatic history the tribe have built up a very complex set of food taboos. These have progressively excluded every food but potatoes. Nowadays they eat exactly 1 tonne per person per year, and their farming techniques will give them 4 tonnes of potatoes from 1 hectare. They need 100 tonnes in all. So they must clear 25 hectares of land.

It's a simple sum, and it applies in the real world too. The procedure is straightforward. To get the total amount of food needed, multiply the population by the amount consumed per person. Divide this total by the yield per hectare – the technology factor. And you know how much forest must be cleared:

farmed area =
population × food consumption per person × area per unit of
food production

Naturally our colonists avoid the steep hillsides and start farming the gentle slopes first. Like most farmers with plenty of land, they shift their fields regularly. They grow potatoes for just one year. Then they move on, clear another patch of 25 hectares, and leave the first one to grow wild again. After nineteen years they have farmed all the area of gentle slopes once over. Now they come back to their first patch. This has now regrown with tall trees. They clear the old patch again.

So far so good.

The island has other advantages. There are no lions, snakes, or malaria. Their isolation means there are no marauding attacks from other islands. So gradually the population grows. One day they count heads, and find that they have become 200.

But they're still using their old technology, still getting 4 tonnes of potatoes per hectare, still eating 1 tonne per person. They measure their potato fields – and discover that they are now farming 50 hectares. The mathematician, now an old man, devises a simple formula that expresses the amount of additional forest they have had to clear.

increase in farmed area =
population increase × increase in food consumption per person × increase in area per unit food production

And, on reflection, he concludes that this formula could be applied anywhere, at any time, to calculate the amount of deforestation that would result from farmland needs.

Of course, even on a remote tropical island, things are never quite this straightforward. There would be several different crops, plus livestock. A second less demanding crop, such as manioc, might be grown in the second year of cultivation of a plot.

These would complicate the calculations, but not the basic relationship. Indeed we can state as a general rule: if technology and food consumption do not change, population growth translates directly into deforestation for agriculture.

When technology lags behind

In the real world technology does not remain static. As we have seen (p. 31ff), growing population density spurs changes in technology to meet the demands of population growth.[6]

But this does not happen at all times and in all places. There are times when food production does not keep up with population growth – and into these abysses great chunks of the human race plunge from time to time.

Ester Boserup, the originator of the theory, herself admits of exceptions. If people had no knowledge of fertilization techniques, for example, increasing population density would lead to a decline in crop yields. If this proceeded far enough, land resources would sometimes become exhausted, leaving people with the stark choice of starvation or migration. Rapid population growth is another situation where, Boserup concedes, adaptation and investment in new techniques might not be able to keep pace.[7]

More generally, the adjustment mechanism does not work –

or is delayed – whenever there are lags in invention and
adoption of new technology. In modern societies thousands of
huge corporations are permanently searching for ways of
advancing the technology frontier so as to improve products or
reduce costs. But in agrarian societies, innovation is sporadic
and jerky. It depends on the insights of individual farmers. Seed
varieties are always being improved: every farmer selects the
best seed for sowing the next year, unless famine forces him to
consume it. On other matters, such as tools and techniques,
there are always trials and errors. And errors usually
outnumber successes, just as a tree produces thousands of seeds
of which only a handful will thrive.

So there are periods of stagnation as well as of rapid change.
The static periods, when technological change does not keep
pace with population growth, may last for decades, and may
involve severe hardship.

Many historians believe that the decades leading up to the
Black Death was one such period. After four decades of
continuous population growth, most good land in Western
Europe was already being cultivated. More and more marginal
lands began to be farmed. Pasture was ploughed up as arable.
Livestock numbers fell and there was less manure. Fallows were
progressively cut back. The result was declining yields, rising
food prices and falling real incomes. Populations in Europe
were already beginning to decline before the Black Death began
in 1348. The plague found easy victims in people weakened by
decades of malnutrition.[8]

And there are blockages to intensification. The medieval field
system, where fields were communally managed, probably
slowed down the introduction of legumes to enrich the soil.
Such a key innovation could not happen before a majority
agreed. Any farmer who planted fodder legumes on his strip
would see them gobbled up by other people's cattle. But under
individual ownership and control, one or two progressive
farmers could start off. Neighbours would follow when they saw
the results.

Most of sub-Saharan Africa has been caught in a similar
technology trap for the past twenty-five years. There are
exceptional physical difficulties to be overcome. The tsetse fly
bans the use of cattle for ploughing and manure across half the

continent. Tsetse flourishes in forest and bush. As population density increases and bush is cleared, the fly retreats. But large livestock cannot be introduced until a much later stage than is possible elsewhere in the world. Irrigation, another major means of intensifying production, is costlier in Africa. Landscapes are flatter, and river flows more variable.[9]

Land tenure is the main institutional block. The development of the state in Africa has raced ahead of the progress of agriculture. In most countries land is still communally owned or, worse still, state-owned. As population density increases, it would be natural for private landownership and a market in land to develop, as it did in Europe – yet the state steps in to prevent it.

Even if we ignore lags and blockages, the Boserup thesis does not mean that growth in yields will always keep pace with growth in population. If that were the rule, there would never be any extension of the cultivated area. There would never be any deforestation for farming.

The theory implies, on the contrary, that deforestation for farming is inevitable. Indeed it is deforestation and shrinking fallows that lead to the problems of soil hardness, weeds and declining fertility, which in turn stimulate agricultural change. Ploughing, weeding and fertilization cost more in labour or cash than slash-and-burn shifting cultivation. Farmers will continue with the latter as long as they possibly can. So long as there is accessible forest land left, farmers will clear it in preference to intensifying on the land they already have. They will intensify radically only when there is no more forest left to clear.

Extension of the farm area, then, comes first. Intensification follows.

World figures on changes in cereal yield and area confirm this. In those parts of the world that still have large land reserves and low population density, a big share of the increase in food production since 1961 was due to area expansion – no less than 51 per cent in Africa. But in the regions with smallest land reserves and highest population densities, most of the production increase came from improvements in yield. Yield increases accounted for 79 per cent of the growth in cereal production in Asia in the last three decades. Between 1983 and 1987 they were responsible for all the production increase.[10]

Growth in the cereal area tends to be fastest in areas of fastest population growth. Growth in yields, by contrast, is slowest where population growth is fastest, and fastest where population growth is slowest. This is almost the reverse of what a simplistic interpretation of the Boserup thesis might predict. But it comes about because population growth is fastest where there are still ample land reserves, and slowest where those reserves are depleted.

So population growth as such does not stimulate yield increase until land shortages begin to develop. And shortages do not start to bite until much of the available forest has been cleared.

Tarmac and concrete

Growing populations need extra land not just for food production and farm livelihoods, but for non-farm land – more space for houses, halls, markets, temples, roads, workshops, bars, and later for factories, offices, shops, gardens, cinemas, swimming pools, restaurants, airports, massage parlours, entertainment arcades.

The area needed per person varies widely from one country to another, but not always in the expected directions. In New York, where homes and offices grow upwards, the ground space used is less than in many Third World countries – as little as 300 square metres per person. In New England, where things sprawl outwards, it may be as much as 2500 square metres. Studies in the Third World have found requirements ranging from 10 up to 800 square metres. The average in one forty-one-country study was 560 square metres per person.[11]

Now towns usually blossom, not in forests, deserts or swamps, but in the heart of agricultural land. Cities, indeed, often swallow up prime farmland in flourishing agricultural zones. Farmland, in turn, has expanded into forests, pastures and wetlands – but mainly forests – to compensate for the loss. Hence urban growth is indirectly responsible for a great deal of deforestation.

New Delhi, for example, is sited on good irrigable land by the Yamuna river. The city spread from 43·3 square kilometres in

1900 to 660 square kilometres at present, engulfing 100 agricultural villages in the process. In just four decades from 1940, Mexico City grew from 117 to 1200 square kilometres. São Paulo expanded from 150 km^2 to 1400 km^2 in the fifty years up to 1980. Between 1973 and 1985 Egypt lost 13 per cent of her farmland to urban sprawl – and Egypt has no land reserves at all where farmers could expand to compensate. In China arable and permanent cropland declined by 7·4 per cent between 1961 and 1988.[12]

The future impact of non-agricultural land requirements will be huge. Taking the average figure of 0.06 hectares per person, the non-farm needs of extra populations expected between now and the year 2150 will take up roughly 3.6 million square kilometres of land in developing countries. This will swallow up the equivalent of almost half the total Third World cropland of 8 million km^2 in 1988. To compensate, cropland will have to expand into forest and other marginal areas.[13]

Assigning blame

The forest – just like wildlife – is suffering from multiple attacks from many directions. Many of these attacks combine to strengthen their impact. While farming and ranching eat at forest from without, logging degrades it from within and opens it up to faster penetration by farmers. Government policies, in turn, have a strong effect on the extent of logging and ranching, and the amount of environmental damage they do.

As the forest area dwindles, farmers add to forest degradation by hunting, grazing animals and gathering fuelwood. At later stages, when countries begin to industrialize and car ownership increases, acid rain and low-level ozone weaken forests further. Global warming poses the biggest threat of all (see pp. 210–6).

It is an interacting system in which population growth, consumption patterns and technology have the primary impact – but are affected by many other factors ranging from government policy and staffing to the form of ownership of the forest.

The decisive role of population growth is clear. A number of surveys have found a strong negative relationship between forest cover and population. Put simply, the more people there are in any given area, the less forest. Finnish forester Matti Palo and colleagues found a very high negative correlation between population density and deforestation. By contrast, there was no significant correlation with the level of industrial wood production. The link with population growth is so strong that the FAO's forestry department is using changes in local population density to predict deforestation.[14]

Local studies confirm the link on the ground. A recent survey of deforestation in Madagascar found that in areas of low population density – mainly in the north of the country – over half the original forest area remained. In areas of high population density all but 19 per cent had been cleared.[15]

We can arrive at some idea of the contribution of population growth to deforestation with the help of FAO's land-use figures.[16]

Between 1973 and 1988 some 1·45 million square kilometres of forest were cleared in developing countries.

Non-agricultural land – for towns, roads, and so on – expanded by 856,000 square kilometres, accounting directly, but mostly indirectly, for 59 per cent of deforestation. This was almost entirely due to population growth.

Farmland expanded by 487,000 square kilometres. Assume that some 29,000 km^2 of this came from drained wetlands, and 58,000 km^2 from pasture in Africa. Then expansion of farmland could have been responsible for some 400,000 km^2 of deforestation. Perhaps 28 per cent of this expansion went towards increased consumption of food and cash crops per person (see appendix, pp. 307–9). So population growth would account for 288,000 km^2 of the deforestation which was due to farming.[17]

Adding this to the non-farm needs produces our final rough estimate: population growth was probably responsible for 1·14 million km^2 of deforestation – around 79 per cent of deforestation between 1973 and 1988, and therefore for 79 per cent of any loss of species due to deforestation.

Some 8 per cent was due to increased consumption of agricultural products.

The remaining 13 per cent was due to ranching in Latin America (p. 95), which has little or nothing to do with population growth.[18]

A threat – and a promise

Our discoveries so far are double edged.

They contain a clear threat. The process of deforestation has so far been an inevitable part of population growth in all parts of the world. Our formula:

increase in farmed area =
population increase × increase in food consumption per person × increase in area per unit food production

has an unavoidable consequence: whenever population increase, multiplied by increased consumption per person, outstrips the growth in yield, the farmed area will increase at the expense of forest and woodland.

But there is a promise, too. When growth in yield equals growth in population and food consumption, deforestation for farming will slow down and may halt.

And we can go further still. When yield growth outpaces its two rivals in the race, the need for farmland will decline. As long as cities are not growing too fast, some reforestation will occur.

This is precisely what has happened in many developed countries and in China.

In very rough terms, the shift from net deforestation to net reforestation came in the 1940s for Europe, the 1950s for North America, around 1970 for Russia and 1980 for China. The precise circumstances varied. Except for North America, overall population growth rates were 1·2 per cent per year or less. And except for China, rural population growth had halted or just turned negative at around the period when deforestation stopped. The Chinese exception may well be due to the very fast rate of growth in crop yields.[19]

At some stage, then, there will be an end to the destruction of the forests. Assuming that a halt in rural population growth is a prerequisite, we might expect net deforestation to halt in South East Asia perhaps by 2005–2010, and in South Asia ten years later. In Africa it could well continue until the middle of the next century.

In South America the total rural population started shrinking after 1975. But the lunatic practice of subsidized ranching has

kept deforestation going longer than is natural. Gross inequalities in landholding, mechanization in agriculture, and
recession due to massive debt burdens, have also closed up
alternative job opportunities, pushing the poor to colonize
rainforest. No one would sweat on poor soils in malarial areas if
they had any alternative. Once these problems are resolved, net
deforestation in Latin America could well halt. Urbanization
could be the saviour of the Amazon.[20]

In most cases, however, respite will not come until massive
damage has been done and a great deal of forest and wildlife
has been lost forever. In some cases – such as western
peninsular Malaysia or Côte d'Ivoire – it does not come until
the destruction of the original rainforest is virtually total. Bio-
diversity and aesthetic value are impoverished for ever.

But the high rainfall in equatorial countries makes it easy to
grow timber or tree crops. Many of the functions of the forest –
timber production, soil conservation, water regulation, climate
regulation – may be restored.

In dryer areas, the prospects for recovery are bleaker. Trees
grow much more slowly. The turnaround may not come until
soil erosion makes the re-establishment of forest difficult or
impossible. The Mediterranean stands as stark warning: many
areas still have not recovered from the depredations of the
classical empires.

Adjusting to wood scarcity

Forest, and tree cover generally, is not merely something that is
cleared for farming. It is a resource in itself, and is managed, or
not managed, just like farmland.

For hunter-gatherers and long-fallow farmers, the forest is the
source of products ranging from honey, fruits, leaves and roots,
to animal protein, fuel, fibres, building materials and medicinal
plants. The shrinking forest loses its capacity to supply these
needs.

But the needs don't disappear. Fruit and vegetables are
grown on the farm. Bees are attracted by artifical hives.
Livestock replaces game.

Wood for timber and fuel is last to shift from gathering to

farming. Long after the last accessible forest is felled, people cut scattered trees for building, and take fallen branches for fuel. But gradually these supplies dwindle too.

Many regions now face fuelwood shortages. In 1980, according to FAO estimates, some 1395 million people in developing countries lived in areas where fuelwood needs could be met only by depleting the stock of trees. Of these, 122 million were suffering acute fuelwood scarcity – that is, they could not meet their minimum needs even by overcutting. In rural areas, where up to 90 per cent of energy is supplied by fuelwood, almost half the population faced fuelwood deficits or scarcity. In Africa the proportion was 58 per cent.[21]

The combination of population growth with widespread overcutting produced a bleak outlook. By the year 2000, the FAO projected, the numbers short of fuelwood would more than double, to 2986 million.

These projections assume that people carry on chopping regardless. But usually they adjust. As wood scarcities develop and free supplies dry up, a market in wood develops. People with land begin to plant trees as a cash crop.

When I visited Kenya in 1980, women in the highlands were spending anything up to three hours a day scouring the landscape for twigs and brushwood. When I returned only six years later, things had changed dramatically. Wood had to be bought on the market. Husbands had to shell out hard cash if they wanted cooked dinners. Planting trees had become a boom industry. Nurseries couldn't keep up with the demand for seedlings. People uprooted self-sown seedlings from roadsides and began to steal them off each other.[22]

A similar adjustment has occurred in neighbouring Rwanda. Old photos from the 1930s show a landscape almost bare of trees – like Ethiopia today. Yet when I visited the country in 1989, every small farm had its own woodlot. Fruit trees such as avocados and mangoes were growing in popularity.

Adjustment does not happen everywhere. As with agriculture, there are blockages. In much of Africa uncertain land and tree tenure, or taboos on women planting trees, stand in the way. In dry areas the slow growth of trees and the relentless pressure of goats makes tree-planting difficult. And so in some places the pessimistic projections do become reality. In the mid

1980s the Ethiopian highlands were virtually bare of trees. Women scoured the fields for cattle droppings and stockpiled dung cakes by the doors of their huts.[23]

Stages in forest management

This process of adjustment to wood scarcity operates at national level too. British geographer Alan Grainger has argued that deforestation may proceed as far as the level at which national wood supplies are threatened. But at that point governments will tend to act to protect forests. This may not occur in some cases until forest area has fallen as low as 0·1 hectares per person. In Japan and Western Europe it reached 0·2–0·3 hectares per person.[24]

In forestry, we can see a similar sequence of phases at work as the transition in agriculture from hunter-gathering, through shifting cultivation, to permanent farming – and the two sequences are related.[25]

Let us call the first stage forest gathering. Population density is low. The forest resource is inexhaustible, and there is no need for any form of management. People simply go in and take what they want, though individual ownership might be asserted over a few prized fruit trees. At this stage the forest area is more or less stable, or declining very slowly. Our Semai friends in the central highlands of Malaysia (see pp. 1–6) are still at this stage.

The second stage can be called forest mining. Population growth speeds up, but farming yields don't keep pace. Farmland and housing eat into the forest area. Fuelwood and timber needs grow in parallel, adding to the rate of deforestation. In most cases this stage coincides with the imposition of state authority. Forest ownership often passes out of local hands into those of the state – but the state lacks the means to control forest use.

At the crisis stage, population growth is rapid, and still outstrips yields. Industrial and urban needs for timber rise sharply. Nations are thoroughly enmeshed in the world trade network: extensive logging for timber adds to population pressures on forests. Deforestation begins to create problems like erosion of farmland, downstream flooding and wood shortage. The state enacts regulations on logging and clearance for farming, but lacks the personnel or the political will to enforce them. The forest resource dwindles rapidly.

The final stage, reached after adjustment to crisis, is one of sustainable use of forests. Population density is high, but the population growth rate is slow. Crop yields are growing faster than population. Urban expansion slows or halts. Electricity and gas replace wood as the chief domestic fuel. Concrete and steel take over as building materials. Privately owned forests expand, and commercial loggers replant knowing they will put themselves out of business if they fail to do so. The state is strong and rich enough to enforce logging regulations in exploited forests, or to police national parks where no logging is allowed. Rising farm yields mean that food requirements can be met from a smaller area. Deforestation slows to a halt.

As leisure demands for forest grow, the forest area grows again. But its rich diversity is lost forever, or confined to a few relics.

8

A STERILE PROMONTORY:
land degradation

Our mountains crumble, and the marsh of Chü-yeh
overflows.

Song of Hu-tzu, c. 109 BC[1]

Almost two and a half millennia ago, Plato identified the links
between deforestation, erosion and the drying up of springs with
surprising accuracy. 'What we now call the plains of Pheleus,'
he wrote of his home territory, Attica,

> were once covered in rich soil, and there was abundant
> timber on the mountains, of which traces may still be seen.
> Some of our mountains at present will only support bees. But
> not so very long ago trees fit for the roofs of vast buildings
> were felled there, and the rafters are still in existence. There
> were also many other lofty cultivated trees which provided
> unlimited fodder for beasts.
>
> The soil got the benefit of the yearly 'water from Zeus'.
> This was not lost, as it is today, by running off a barren
> ground to the sea. A plentiful supply was received into the
> soil and stored up in the layers of clay. The moisture
> absorbed in the higher regions percolated to the hollows, and
> so all quarters were lavishly provided with springs and rivers.
> To this day the sanctuaries at their former sources survive.
>
> By comparison with the original territory, what is left now
> is like the skeleton of a body wasted by disease. The rich, soft
> soil has been carried off. Only the bare framework of the
> district is left.[2]

His comments could have been written about the Sahel, the
Himalayas or the Andes today.

Ever since temperatures cooled enough for water to condense, erosion has been part of the earth's dynamics. The tectonic machine thrusts up mountains and volcanoes. The climate machine does its best to level them flat and wash the debris into the sea.

This cycle of creation, destruction and re-creation is a natural process of ceaseless change. It becomes land degradation only in relation to living organisms that seek to survive on it. Land degradation is a significant reduction in the capacity of land to support biological production.

Serious land degradation caused by humans is as old as the mastery of fire. The introduction of agriculture widened its scope – and aggravated its consequences. In Mesopotamia it took the form of a build up of salts in irrigated areas. Progressive stretches of the lower Tigris and Euphrates had to be abandoned, and the focus of each successive empire moved northwards. The Maya civilizations of Central America may have fallen victim to erosion caused by deforestation.

Plato's Attica had a thin and vulnerable soil. As deforestation and soil erosion proceeded on her hinterland, Athens could no longer meet her food needs. As early as 594 BC the legislator Solon banned the export of corn. The only crops that would grow well on her exhausted soils were olives and vines, which tap subsoil layers, and fruit best when somewhat starved of nitrogen. So Athens exported these, and imported corn and timber. Her ecology forced her to build the greatest trading empire of the Eastern Mediterranean – an expansion that inevitably brought her into conflict with Sparta. It could be argued that deforestation and erosion in Attica was one of the major causes of the Peloponnesian war.[3]

Degradation and desertification

Today land degradation is global in extent. It occurs in developed countries, where it may be masked for decades by rising applications of fertilizer. In tropical countries, with their scouring rains, it is more serious still.

The United Nations Environment Programme has estimated that 6–7 million hectares of cropland are being lost worldwide each year through soil erosion, and another 1·5 million of irrigated and wetland to salinization or waterlogging. These losses would represent just under half of 1 per cent of the world's cropland in 1988, and about two-thirds of a per cent of the irrigated area.[4]

More detailed assessments are available only on a regional basis. In Africa, for example, 16 per cent of the land area is at risk of moderate or severe water erosion, and twice that proportion to wind erosion. In the Near East the proportions are slightly higher – 17 per cent and 35 per cent respectively.[5] In Latin America, 14 per cent of the land area has already been affected, half of that severely.[6]

Agronomist Harold Dregne has estimated that some 44 per cent of the land area of Asia has lost more than 10 per cent of its productive potential. The proportion affected in Africa was 40 per cent, and in South America 27 per cent.[7]

Desertification is the most widely publicized form of degradation. The word evokes nightmares of inexorably advancing deserts. But few places match this emotive image. Like tides on the beach, deserts retreat and advance from one year to the next. In Africa desert vegetation zones on the Sudan–Chad borders moved 200 kilometres south in 1984. The following year they moved the same distance back northward. Desertification spreads far beyond the borders of the desert proper, eating away inside productive areas, like moth holes in clothing.[8]

Desertification was defined at the 1977 United Nations Conference on the topic, as 'the diminution or destruction of the biological potential of the land', which could 'lead ultimately to desert-like conditions.' Thus it could be applied not only in the rather unusual case of desert frontiers advancing – but to any non-humid area undergoing degradation.

Some 61 per cent of the world's productive drylands were moderately or severely desertified in 1985, according to UNEP. In developing countries the proportion was as high as 77 per cent – including 80 per cent of rangelands, 75 per cent of rainfed cropland and 35 per cent of irrigated land. About 10 per cent of the rural population of developing countries lives in affected areas – as many as 270 million people in 1990.[9, 10]

These estimates are rough guides rather than precise

measurements. In many cases they are little better than expert guesses. And the definition of desertification is so broad that it includes almost every type and degree of land degradation.[11]

The term desertification is perhaps best reserved for processes that could lead to a complete loss of productivity – for severe crusting, erosion to gullies or bedrock, formation of impenetrable hardpans, invasion of sand dunes. Lesotho has 7000 gullies covering 60,000 hectares, about one-ninth of the crop area. They snake down through cropland from the edges of weathered sandstone outcrops. Much of central Madagascar is scarred by ugly *lavakas* – circular collapses of half hillsides.

Sand invasion is serious in areas that border deserts. In Niger the northern limits of cultivation are under continous assault. Sand buries millet seedlings. It heaps like snow drifts against walls and fences. A line of shrubs can foster a baby dune, which grows to the size of a great humped Leviathan. And there is one of these monsters every kilometre or so.[12]

In Egypt, invasion of desert sand has damaged a strip 3 km wide stretching for 1200 kilometres on both sides of the Nile valley. As much as a third of Egypt's cultivated area is affected, and a quarter of agricultural production has been lost.[13]

Rangeland is badly affected by desertification. Conditions in most areas are poor, sometimes alarming. Rangeland in seventeen out of thirty-two countries or regions analysed by the World Resources Institute in 1986 came into the high-risk category. Sixteen of these were in the developing world. Even in the United States, only one-third of rangeland was considered in good condition; 40 per cent was poor or very poor.[14]

Excessive livestock pressure alters the ecology of ranges. Palatable shrubs are weakened, forced to shift resources from root growth into production of leaves, to replace what has been eaten. Eventually the plant dies. Plants that livestock dislike gain an advantage, and spread. Soil is compacted, the cover of vegetation thins, and soil erosion begins. On Syria's rangeland the original plant community has been eradicated. In Iraq it has been mainly replaced by unpalatable shrubs. In Iran most palatable perennial species have disappeared.[15]

Some experts stress that we know too little about the biology of rangelands to say that they are deteriorating. Vegetation boundaries change from year to year according to rainfall. Dry rangeland areas are also surprisingly resilient. When left to rest for a few years, windblown seeds root and collect soil around themselves, and the range blooms again. When the degraded Simbol area in Kenya's Pokot district was closed off to herds because of armed raiding, it became impenetrable bush.[16]

The toll of erosion

But desertification proper is only the most visible tip of a huge iceberg. Arable land all over the world is degrading: losing fertility, depth, structure.

Soil erosion is the world's most widespread form of land degradation. In the United States, cropland lost an average of 10·5 tonnes of soil per hectare each year in the 1980s, and 44 per cent of cropland was losing soil faster than it forms. In developing countries, rainfall is torrential, and erosion hazards are more severe. Losses of over 100 tonnes per hectare have been reported from parts of India, China, Yemen, El Salvador, the Dominican Republic, Kenya, Madagascar and Ethiopia. In India more than half the cropland has been degraded to some extent by soil erosion.[17]

Soil erosion is a problem both for farmers and for downstream dwellers. It depresses yields. US corn and soybean yields were 2.5 per cent lower in 1980 than they would have been without the erosion that occurred over the previous thirty years. In tropical soils, nutrients are often concentrated close to the surface, and the effect can be much more dramatic. On a test plot at the International Institute of Tropical Agriculture (IITA) in Nigeria, maize yields plummeted from 2 tonnes to 0·7 tonnes when the top 10 centimetres of soil were removed. When 20 centimetres were removed, grain yield dropped to only 0·2 tonnes.[18]

Fertilizers can compensate for erosion to some extent, but usually by increasing other environmental impacts such as water pollution. Erosion and loss of organic matter greatly reduce the effectiveness of fertilizer. In one IITA experiment using high levels of fertilizer, maize yields fell by 55 per cent when 10 centimetres of topsoil were removed.[19]

The Food and Agriculture Organization has tried to assess the global damage that unchecked land degradation could do. If no soil conservation measures were undertaken, some 544 million hectares would be lost. This is 18 per cent of the total potential cropland. Erosion would also cut yields on the remaining land. The total loss of productivity from rainfed cropland would be as high as 29 per cent.[20]

Erosion has serious effects off the farm too. Streams are clouded. Fish production declines. Water supplies are degraded. River navigability is reduced. Flood risks increase. In the United States some 1·5 billion tonnes of silt – 10 per cent of river-sediment discharge – is deposited in dams and reservoirs every year, cutting their effectiveness and lifespan. The total cost of off-farm damage in the United States was estimated at $6 billion in 1980.[21]

The world's 229 million hectares of irrigated land do not suffer heavy erosion. But they have problems of their own. Waterlogging occurs when crops are overwatered. The excess sinks into the soil, the water table rises and chokes plant roots. In some irrigation schemes, water tables have risen by anything from 30 centimetres to 3 metres per year.[22]

When water tables rise to within 4 or 5 feet of the surface, another widespread problem, salinization, appears. Then evaporation leaves the soil salter than before. This depresses yields, and can eventually poison the land for crops. Between one-third and one-half of the world's irrigated lands are thought to be affected to some degree by waterlogging, salinity or alkalinity. The share affected in the arc between Iraq and India ranges from 21 per cent to 34 per cent. As many as 1·5 million hectares of irrigated land may be lost to salinization each year.[23]

Loss of soil nutrients is equally serious. Crops rob the soil: every tonne of rice removes 20 kilos of nitrogen, 11 kg of potassium and 30 kg of phosphorus. Other trace elements are also depleted, including sulphur, zinc, iron, manganese and boron.[24]

A recent FAO study warned that African farmers were mining soil nutrients at such a rate that serious declines in crop production were inevitable within the next ten years. In sub-Saharan African countries 22 kilos of nitrogen were being drained from each hectare every year. In Malawi, Rwanda, Kenya and Ethiopia losses were twice the average, or more.[25]

Even where fertilizer is applied, soil fertility can still decline if the amount used does not restore losses. Asian farmers apply almost six times more fertilizer per hectare than Africans. Yet nitrogen deficiency is almost universal in Asia. And the fertilizers that are applied usually have too high a proportion of nitrogen. The result is that crops mine the soil of all the other nutrients. Phophorus deficiency affects 73 per cent of farmland in China and 80 per cent in Pakistan. Zinc deficiency has reached epidemic proportions in many countries. Sulphur is deficient in 70–80 per cent of soils in Bangladesh and on one-third of Chinese cropland.[26]

Looking for the causes

One hard-nosed author has claimed that soil erosion could be beneficial: it removes soil and water from hilly sites, where their productivity is poor, and delivers them to lowland sites where they are used more efficiently. It is true that the fertility of the best alluvial farmland in the world depends on soil erosion from upstream. However, one man's gain is another's loss: the transfer of soil wealth pauperizes hill dwellers.[27]

Critics of the radical right, like American economist Julian Simon, have tried to deny that land degradation is occurring on any significant scale. If it were, he says, there should be a decline in arable land. In fact, the figures show a continuing increase almost everywhere. Even if the world were losing cropland at the rate of 6 or 7 million hectares each year, there was still a net increase in total cropland of 2·66 million hectares each year in the 1980s, after any losses to degradation.[28]

But this ignores vital questions of land quality and fertility. Much urban expansion occurs in the very best agricultural land. Compensating expansion is usually into land of lower quality. Fertilizer use masks falling fertility.

Evidence from the ground, rather than from statistics, shows that land degradation is a severe problem throughout the world. It is regarded as a major threat not just by environmentalists but by governments, agriculturalists, and increasingly by farmers themselves. The claim that it is not serious does not stand up to detailed investigation.

The left analysis has taken a different stance. It accepts for the most part that land degradation is a problem, even a serious one. But it denies that population growth is a principal cause.

'In Africa,' US food and agriculture writer Frances Moore Lappé asserts, 'it is colonialism's cash crops and their continuing legacy, not the pressure of population, that are destroying soil resources.' Colonialists expropriated much of the best land for plantations, and grew continuous cash crops for export. Displaced local farmers were pushed on to marginal land. Throughout the world neo-colonial élites still monopolize the most fertile land. 'Soil erosion occurs largely because fertile land is monopolized by a few, forcing the majority of farmers to overuse vulnerable soils.' Lappé accepts that population growth can 'exacerbate' environmental degradation, but denies that it is a 'root cause'.[29]

British erosion expert Piers Blaikie offers a more sophisticated approach. Land degradation, he says, happens when surplus is extracted from poor farmers. This extraction can take many forms: rents, excessive taxes, underpayment for produce. The cultivator, at risk of falling below subsistence level, is then 'forced' to extract surplus from the soil simply to survive. The only way to do this is to mine its stored fertility, trees or other biological capital. This marginalization is therefore a primary cause of land degradation. Blaikie accepts that population pressure can create stress, but assigns little weight to it in the balance of his analysis.[30]

The cash crops myth

Cash crops are universal Aunt Sallies. They are blamed for causing environmental degradation, famine and child malnutrition, for pauperizing peasants, forcing the poor into marginal areas.

It does, on the face of it, seem obscene when regions where thousands or millions starve produce crops for export, or when drought-hit areas continue to export cotton. But the reaction is emotional rather than rational. Attacking cash crops is rather

like attacking bank accounts because they are sometimes used
by drug dealers. The real culprits for poverty, malnutrition, and
marginalization are deeper structural factors: unequal landownership, oppression of women, low agricultural prices.
Where these apply, the rural poor will suffer, whether cash
crops or food crops are grown. Where they are reversed, the
poor will fare better, whether they grow food crops or cash
crops. To the landless Brazilian labourer starving outside the
barbed wire of vast *latifundia*, it makes little difference whether
the landlord next door is growing pineapples or wheat. Either
way he will get the mouse's share of the benefits.

It's worth examining the facts in more detail. Two cash crops
peanuts and cotton   have been put in the dock for Africa's
repeated famines. The argument is simple: these export crops
displaced food crops for local consumption. That is certainly
true of the colonial period. But in recent decades, with two
severe famines in twelve years, the reverse has happened: food
has been displacing cash crops. The combined area under
peanuts and cotton in Africa actually decreased by 2 million
hectares between 1970 and 1984, from 9·1 to 7·1 million
hectares. Over this same period the area under the Sahel's two
main grain crops, millet and sorghum, increased by 2·4 million
hectares. Sahelian farmers adjusted to poor prices and increasing population pressure by shifting away from cash crops into
subsistence food crops.[31]

The total area under Africa's nine main cash crops decreased
by 7 per cent over the fourteen years from 1970, while the grain
area increased by 18 per cent. By 1983–5 cash crops occupied
only 12 per cent of Africa's arable area.

Cash crops certainly got the lion's share of government
credit, inputs and extension advice. Food crops would have
prospered better had they got the same attention. But cash
crops did not perform better than food crops, as you might
expect if the conventional analysis were true. On the contrary,
over the fifteen years from 1970, the yields of all grain crops
rose, but the yields of the top three cash crops in terms of area
fell. African countries divided into two main groups: a large one
where cash-crop and food-crop production per person both
declined, and a much smaller group where both rose. Other
factors, such as government pricing and marketing policies, or

the incidence of war, accounted for the success or failure of both.

Cash crops, just like food crops, vary a great deal in their ecological and social effects. There are some situations where they wreak havoc. Sugar-cane, with its high demands for nutrients and seasonal labour, can damage soils and leave the landless unemployed for half the year. I have met Kenyan farmers, near the sugar factory in Mumias, who hired their land out on long-term contracts for intensive sugar production, and deeply regret it. Their trees were felled, their strips of thatch-grass ploughed up. They ended up short of cash for most of the year, having to buy fuelwood, and being trucked around at harvest time to chop cane.

Yet a few kilometres away I met other smallholders merrily growing coffee bushes, intercropped with tomatoes and cabbages, as part of a very diversified mix. The coffee helps to provide cash for kerosene, salt, radios, bicycles, and so on. The world economic system rarely guarantees a fair or a stable price. Yet the prices are often more attractive than those for grain. With the cash earnings the farmer, and the country, can buy as much or more maize than would have grown on the same area – and have cash left over for other things. When farmers are free to decide which crops to grow, in the light of world market prices, they will choose the crops that suit them best. And sometimes these will be cash crops.

Just as important from the farmer's point of view, cash crops spread risks. Farmers who grow only food crops are more likely to suffer catastrophic crop failure. An invasion of stem borers that ruins maize will leave coffee bushes untouched. Cotton may survive a dry spell that kills sorghum. Cash crops have different harvesting and planting times. They spread labour needs and cash income through the year.

But it is important that farmers should be genuinely free to decide for themselves what crops to grow, and that they should receive the full market price for their produce, preferably spread through the year rather than in a single lump sum. There should be no compulsion to grow cash crops; no oppressive long-term contracts with factories or buyers; no centrally determined cropping schemes in large agricultural projects; and no artificial encouragement of crops that risk oversupplying the world market. But these provisos apply to food crops too. Food crops grown for sale are also cash crops.

The environmental effects of cash crops vary from one crop to another, and from one technique to another. Tobacco, which needs a lot of wood for curing, may accelerate deforestation. Some fertilizer formulae for cotton acidify soils and deplete other nutrients. But unbalanced fertilizer inputs for food crops do exactly the same.

Perennial tree and shrub crops like cocoa, coffee, tea, and fruit, can protect soils as part of agroforestry systems. Some food crops can degrade soils rapidly unless conservational techniques are used. In Rwanda, for example, tobacco tops the list of erosion-creating crops. But the food crop sorghum comes a close second, with twice as much soil loss per hectare as the cash crop pyrethrum. Maize comes third. Even when inter-cropped with beans it creates fifteen times more erosion than coffee. Coffee bushes are usually mulched in Rwanda. Coffee was the best soil conserver out of forty-one crops and combinations measured.[32]

By contrast: Ethiopians have a tiny grain, *teff*, out of which they make big brown rubbery pancakes. *Teff* is traditional, it is food. It has nothing to do with all those evil forces we are told create soil erosion. But to get the superfine tilth *teff* needs, farmers lash their plough teams across the field two or three times over. This leads to haemorrhagic soil losses: up to 170 tonnes a year from each hectare, among the very worst in the world.

As in every other sphere we are dealing with, the problems are diverse and complex. There are no simple slogan-type answers like 'cash crops are always bad', or 'cash crops are always good'. Some crops, some cultivation methods, some decision and marketing systems are harmful. Others are beneficial. And this applies to food crops as well as cash crops.

9

A LITTLE PATCH OF GROUND:
living on the margin

> Nor is it sensible to believe that the Earth grows old
> like a mortal man. Such misfortunes do not come
> upon us through the fury of the elements, but
> through our own failings.
>
> Columella, first century AD[1]

Let us move on to our next suspect in land degradation:
marginalization. 'Marginal' means, literally, close to the edge.
Marginal people live on the edge of the economy, and often on
the brink of survival. In 1980, almost three in five of all rural
households in developing countries – some 132 million families
– had holdings too small for subsistence. Another 31 million
families – 13 per cent of the total – were completely landless.[2]

Marginal areas lie on the periphery of prosperous zones.
They are prone to rapid degradation when their natural
vegetation is disturbed. These regions are vast. In the FAO's
study on the population-carrying capacity of land in developing
countries, a massive 2453 million hectares – 38 per cent of the
total area – could not support their populations using low levels
of farming inputs. And these are, generally, areas where low
inputs predominate. More than a third of the land area of the
Far East was involved, almost half of Africa, and three-quarters
of the Near East.[3]

Most of the critical zones were marginal. Some were
mountainous – the spine of Latin America; the East African
highlands; the ranges between Turkey and Afghanistan; and the
mountain regions of Indo-China. Others were semi-arid –
north-east Brazil; the Sahel and the Horn of Africa; a dry
corridor running across Southern Africa from Namibia to
southern Mozambique; the Yemen; and the Deccan in India.

In the Near East, Central and South America these zones could produce enough food on a long-term basis for only one-third as many people as were living there in 1975. In Africa they could feed two-fifths, and in the Far East three-fifths. By 2025 the populations will have at least doubled or trebled.

Marginal areas are more prone to degradation. On steeper slopes the erosion risk is far higher. In semi-arid areas vegetation, once lost, is harder to re-establish. Rainforests, not usually thought of as marginal, often have problematic soils that degrade quickly when the forest is removed.

Marginal areas need conservation *more* than non-marginal areas. Yet the basic productivity of the soil is low. Subsistence farmers may not worry about this as long as they can grow enough to survive. But once an area is meshed in national and international markets, its low productivity counts against it. Marginal land is less valuable than non-marginal land – but the costs of conserving it are higher. In a market economy, marginal land is least likely to be conserved. When it is severely degraded, it may simply be abandoned.

There is an unholy marriage between marginal areas and marginal people. One assessment made in 1989 suggested that some 370 million absolute poor lived in marginal or fragile lands.[4]

Marginal areas marginalize their inhabitants. At the best of times they provide a poor subsistence. Semi-arid areas or problem soils will never give as good a living as well-watered plains, volcanic or alluvial soils.

Yet the poverty of the people further impoverishes the area. They may turn to exporting their natural capital. The dry zones of East Africa, even more than the leafy highlands, need their trees. But it is the dry areas that harvest their wood to make charcoal. It is one of the few assets they have to sell.

Marginal people and marginal environments are chained together and become the agents of each other's destruction.

A dwindling inheritance

People can become marginalized even within areas that are basically prosperous. The Malthusian school see rapid population growth as the mechanism that brings this about. In most of the Third World, when someone dies, land is divided equally among the children – usually only the males. Where children outnumber parents the average size of holding declines steadily. After repeated division the smallest holdings barely provide a subsistence. Anything that pushes the owner below that level – serious illness or accident, drought, pests or disaster – can force him to sell. Or he may go into debt, mortgage his land as collateral, and lose it if he cannot repay.

For the radical school, exploitation and expropriation are the chief causes of marginalization. Sometimes the poor are deprived of their land by violence or fraud. Sometimes commercialization gives an advantage to large farmers, who have better access to markets and can afford inputs like fertilizer and high-yielding varieties. As higher yields increase production and push crop prices down, small farmers lose income and may be driven into debt. Eventually they sell out to large owners, or lose their land through mortgage foreclosures.

In reality, these two accounts are not alternatives. Both are needed for a complete view. Population growth and inequality work together in destructive synergy. Population pressure is almost always a crucial factor in marginalizing poor farmers. But its consequences are aggravated – again almost universally – by inequality in landholding and in access to inputs, credit, markets, legal institutions and political influence.

The rival schools agree on some points. Landlessness will increase with time. The average size of holdings will fall. Smallholders will do most selling of land, and large landowners most buying. But there are predictions on which they differ, and these can be used to test their validity. The Marxian theory predicts that landholding will become increasing polarized: the big will get bigger and the small will get smaller. But if the Malthusian theory is correct, large farms will shrink too.[5]

The evidence from Asia and Africa tends to support the view that population growth is often the dominant factor. In India the average landholding dropped from 2·28 hectares in 1970–1 to 1·68 hectares in 1985–6. Marginal farms with less than 1 hectare rose from 51 per cent to 58 per cent of all farms. But the big farms did not grow fat by swallowing up the small. Those bigger than 10 hectares saw their share of total farm area decrease substantially, from 31 per cent to 20·5. Shrinkage by equal inheritance worked on rich as well as poor.[6]

Rich and poor are not eternally fixed divisions. There is upward mobility among the poorest, and even more downward mobility among the rich. One study in Maharashtra traced the landownership patterns of individual families over half a century. A quarter of the families who were landless in 1920 had acquired land by 1970, while 44 per cent of those who possessed land in 1920 had lost it fifty years later.[7]

In Pakistan, similar processes were at work. Between 1950 and 1972 the total farm area grew by 39 per cent. But the number of farmers doubled. So the average farm size shrank by 30 per cent, from 3·8 to 2·7 hectares.* The rich in land did not get richer. In 1950 landowners with more than 40 hectares controlled 32 per cent of the area: in 1972 only 17 per cent.[8]

Bangladesh followed a different pattern. Here polarization did increase. In 1960 the bottom 60 per cent of landowners had 25 per cent of the farmland. By 1978 their share had dropped to 9 per cent. Meanwhile the top 10 per cent of landowners increased their share of the area dramatically, from 36 per cent to 52 per cent. Thus 16 per cent of the total farmland passed from·the poor to the rich. But this was a very special case. In the early 1970s a sequence of catastrophes – cyclone, civil war, flood and drought – cut incomes and compelled many smallholders to sell their land simply to survive. Disaster speeded up the rate at which the smallest go to the wall.[9]

Given a short respite from disaster, the engine of population growth in Bangladesh resumed its work, relentlessly shrinking average holdings in all groups. Between 1977 and 1983–4 the average farm size shrank from 1·4 to 0·92 hectares. But this time the rich were not picking up the pieces. The share of the total area held by farms with more than 3 hectares dropped from 32 per cent to 26 per cent.[10]

It is population growth, compounded by disaster, that has marginalized the overwhelming majority in Bangladesh. Once marginalized, they become more vulnerable to dispossession by the wealthy. The pressure is now so great that even a

* One hectare equals 2.5 acres.

completely equal division of farmland among all rural households would create farms averaging only two-thirds of a hectare
each. Within one generation these would be reduced to a
quarter of a hectare.[11]

In Africa population growth rates are so high, that in densely
settled areas a family can jump from smallholding to landlessness in just two generations. In Rwanda the average small-
holder was farming a mere 1·2 hectares in 1984. Since sons
inherit, and each family has an average of four sons, each son
will receive on average around 0·3 hectares. The grandsons, at
projected future fertility rates, would receive less than 0·1
hectares each, some time around the year 2040. Thus in the
space of only sixty years the size of the average landholding
would have dwindled by more than 90 per cent because of
population growth alone.

Alphonse Njagu farms a third of a hectare on the shores of
Lake Kivu in Rwanda, under the louring cinder cone of the
volcano Nyiragongo. His father had a hectare of land – just
about enough to survive on. But on his death this was divided
between three sons. Alphonse in turn has three sons aged
between sixteen and twenty. I asked him if he would give them
any land when they married. 'I can't even give them a plot to
build a house,' he said, shaking his head. 'They'll have to look
after themselves.'

Forced out – or squeezed out?

People also become marginalized when they are pushed into
marginal land.

The radical school emphasizes expropriation again. The poor
get driven out of better areas by colonial powers, rich
landowners, multinational food companies, cash-crop and dam
projects. They head for the only land that's free – and that is
usually marginal land.

Malthusians counter: it is population growth in the better-
favoured areas that fills up land there, so people who want land
are obliged to move into marginal areas.

As usual the situation is diverse. In some parts of the world
expropriation has undoubtedly been the dominant factor in the

past. In Latin America the Spanish and Portuguese seized the best land. Indians were forced to become labourers, with at most a small infertile plot. Where the Indians died out, Africans were imported. Conquest and slavery created a large class of marginal and landless peasants.

The colonial stamp moulded Latin America's present-day pattern of landholding. The largest 8 per cent of holdings occupy 80 per cent of farmland. The smallest 66 per cent of farms are squeezed into just 4 per cent of the area. The poor who cannot find work locally migrate to cities, or farm hillsides or forest clearings that may only provide a few years' harvests before they become useless for agriculture.[12]

Colonialism and its aftermath expropriated black people on a vast scale in Eastern and Southern Africa. In Zimbabwe the 1969 Land Tenure Act gave 16 million hectares of the best land to just 6700 European farmers. One hundred times that number of black farmers had to make do with the same amount of poor land, three-quarters of it marginal. Today the black areas are deforested, overgrazed, overcropped, eroded.

In South Africa, whites had 87 per cent of the land in 1980, though they made up only 13 per cent of the population. With blacks the ratios were exactly reversed. Each white had access, on average, to forty-five times more land than the average black. Not surprisingly pressure on the land in black areas is intense. Almost all South African whites have electricity. But four out of five South African blacks have no supply, and have to overcut trees to get fuel.[13]

Expropriation continues today throughout the Third World, through fraud, for land settlement and for large scale development projects such as dams. The Narmada dams in India may displace up to one million people.[14]

Yet even where expropriation is the main driving force, the impact of population growth is powerful. For it is through population that inequality and expropriation work their impact on the environment. They confine the oppressed to a smaller area, and artificially boost population density. Natural population growth goes on to increase that density, and worsens the problem.

In most countries population growth is now the main factor pushing people into marginal areas. As valley bottoms fill up, people are compelled to move uphill, and farm steeper slopes which are easily degraded. This gradual climb can be seen all over the developing world, whatever the form or distribution of land tenure. It occurs where land ownership is grossly unequal, as in Guatemala. It happens under socialist ownership of land, as in Harerghe, Ethiopia. And it is found among masses of fairly equal smallholders, from Machakos in Kenya to the slopes of Java's volcanoes.

Population pressure pushes people into dryer zones too. In the Sahel the margin of cultivation has shifted north, into areas suited only for pasture. In East Africa the move is downhill, from the well-watered lush green hills, to the desiccated tawny plains. In Kenya the fastest population growth has been in the dryest provinces. In Rwanda, population pressure in the highlands has pushed people into the semi-arid lowlands, accelerating deforestation and desertification.[15]

The rangelands of Maasai pastoralists have been squeezed from two directions. Their own populations have increased. Under the Ngong hills near Nairobi I met one patriarch whose thirty-five living children, by five wives, filled up the local primary school. And, as land became scarce in the highlands, people have started farming the rangeland. In the 1880s, 45,000 Maasai occupied 200,000 square kilometres of range. By 1961 their territory had shrunk to 93,000 km^2, yet their numbers had swollen to 117,000.[16]

To pay five ducats, five, I would not farm it

Lesotho is a marginal area peopled by marginal people.

The Kingdom is surrounded on all sides by South Africa. From the air, the border can be seen traced in soil: healthy browns and greens on the South Africa side, shades of pink, yellow and white on the Lesotho side.

Between outcrops of old sandstone that still carry the footprints of dinosaurs, the lowland farms are rifted by fierce gullies as jagged as forked lightning. Each year Lesotho loses 40 million tonnes of topsoil, 18–20 tonnes from every hectare. At that rate, the Kingdom will be eroded to bedrock in fifty years time. Almost two-fifths of cropland and three-fifths of rangeland have erosion rates three times higher than the rate at which topsoil is formed.[17]

Two-thirds of the area is pasture, steep highlands carved out of basalt, with sharp peaks that the Basotho call Khuahlamba – the range of pointing spears. In many parts the wiry grass has been eaten almost bare, scarred with cattle trails and landslips. Thirty per cent of the rangeland has lost more than half its cover of vegetation.[18]

When we follow the chain of causes behind Lesotho's marginalization, we are led in a dozen directions, which map out the full complexity of land degradation.

Begin at any point you like in the vicious circle. Population density plays its part. At first this density was the artificial creation of violent expropriation. The Basotho were driven into the hills by Shaka Zulu. Then the Boers robbed them of their most fertile lands in the middle of the nineteenth century. Since then natural growth has compounded the problem. The population rose from 734,000 in 1950 to 1,774,000 in 1990.

Lesotho is almost totally bare of trees. All dung is burned as fuel. Crop residues are also burned, or eaten by livestock. Nothing of what is taken out of the land is returned to it. Deep in the hills you can get beer anywhere – on an upcountry trip I counted one empty can every three metres – but chemical fertilizer is unobtainable outside pilot agricultural projects.

No fertilization means low yields. Maize is the chief crop, but Lesotho farmers get only 750 kilos per hectare – less than half the low African average. Low yields mean that the cover of leaves protecting the soil is thin. Lack of organic matter leads to lower humus levels and higher erodibility.

Livestock here are part of the problem rather than part of the solution. During the summer months, most cattle are kept in the highlands pastures, tended by herd boys aged as young as seven, living a brutal life in isolated huts. Although village territories are recognized and defended by the herd boys, the rangeland is state-owned. No one has any real incentive to limit the number of their cattle to suit range conditions. Lesotho has three times as many livestock as the rangelands can sustainably carry. Cattle in Lesotho are not simply livestock. They are savings stored on the hoof, walking brideprice, living larders awaiting slaughter at funerals and circumcision ceremonies.

As the cold winter approaches, the cattle are brought down to lowland farming areas where they graze on crop stubble, speed soil erosion, and hamper conservation efforts. All cropland in Lesotho is state-owned and communally controlled. After the harvest, fields are thrown open to village livestock to eat the stubble. There would be no point building terraces: they would be trampled; or planting trees or hedges: the seedlings would be eaten. The effect of communal land tenure is inscribed on the landscape. In the gardens around houses, where people have full security, they have planted hedges, fences, fruit trees, vegetable plots. But the communal fields are bare and windswept.

Communal land is allocated by hereditary chiefs, often in exchange for bribes. Not surprisingly the chiefs have resisted plans to strip them of their powers. Owners of large livestock herds, too, want to keep the right to graze their animals on the fields of others. Almost half the population in Lesotho have no cattle. They suffer communal grazing on their fields without deriving any benefits. The land tenure system is a tyranny of chiefs and cattle owners over the rest.

South Africa has a big impact on soil erosion here, as well as in its own homelands. In the 1980s, 120,000 men in their prime were away from Lesotho at any one time, most of them in the Republic's mines. Most farming is left to the women, who divide their limited time between home and field.

I met Leche Matsolong resting on her hoe, in a field of about one hectare. A twenty-foot-deep gully was gradually swallowing her land. Her maize plants were sparse. Some had been grazed by neighbours' cattle. Two-thirds of her plot was smothered in purple and white cosmos flowers – the most common weed in Lesotho.

Her husband had been in the mines for fifteen years. On home visits he considers himself on holiday and refuses to lift a finger. Leche has five children, three in school, two in the hills as herd boys. None are available to help with her long list of chores. Every day she spends three hours gathering shrubs and dung for fuel, and an hour fetching water. Three days a month she treks to a distant mill to get her maize ground. Farm work has to be fitted in around these other unavoidable chores. Even then it is often interrupted by frequent bouts of illness.

Despite her abysmal maize yields, the family does not starve. Leche's husband brings home cheap South African maize, bought with his mine wages. Mine earnings make up more than 70 per cent of earnings in rural areas. But they are a mixed blessing. On balance they damage agriculture. Farming is pin-money compared with wages. It is not crucial to survival.

Low farm prices do further damage. Lesotho has a customs union with South Africa. But maize costs much less to produce in South Africa than in Lesotho. Local crops cannot compete with cheaper imports. And the cheapness of South African coal works against planting trees for fuel.

The direct impact of population growth

We cannot assign population a precise share in land degrada-tion. There are no clear statistical links of the kind that have been found with deforestation. Soil erosion involves so many factors, from the smallest details of farm implements, through crop patterns, to soil type, that it would be practically impossible to document the population link. The results would vary from one field to the next.

But we can identify the mechanisms by which population growth affects soil erosion. Soil erosion rates depend on several factors. One is the erosivity of rainfall – how much rain falls, how fast, and how big the raindrops are. Thirty millimetres of rain, smashing down in big drops in a cloudburst of ten or twenty minutes, creates many times more erosion than the same amount falling in a fine slow drizzle over the course of a few days.

Steep slopes are much more liable to erosion than gentle slopes or plains, since rain scours off at much faster speeds. Erosion rises exponentially as the slope increases – a slope of one in twelve, over a mere five metres, can produce as much erosion as a slope of one in fifty over 300 metres. Soil erosion also increases with the length of each clear run of slope, as rain running downhill gathers more mass and speed.[19]

Plant cover affects erosion in two ways. It breaks the impetus of raindrops and softens their impact on the soil. Water filters in slowly instead of streaming off destructively. And plants break down to form humus. This binds particles of soil together and makes them harder to wash or blow away. The less plant cover, and the less humus in the soil, the greater the erosion.

Population growth affects all these factors – except the erosivity of rainfall – in the direction of increasing erosion. As population grows, more land is cleared for farming, less is left as forest or fallow. The plant cover of a field of crops is nearly always thinner than forest or grassland with thickly matted roots. Hence any increase in the cropped area usually means a reduction in the vegetation cover and an increase in erosion. And there is less plant material to rot down into humus and hold the soil togther.

As more land is cleared, the clear run of each slope increases. And as the flatter land is occupied, people move on to steeper and steeper slopes.[20]

Four into one won't go

Increasingly, cropland alone has to produce four basic necessities for growing populations: food, fodder, fuel and fertilizer.

Humans are omnivores. Whenever they can afford to they eat meat and dairy produce as well as grain. And livestock provide other services, from ploughing and pulling carts, to providing manure. Hence when people increase, livestock increase in parallel, though usually more slowly. In 1989 there were around 1280 million cattle – roughly one for every four humans – and 1702 million sheep and goats. Between 1970 and 1989 the cattle population grew at an average of 0·8 per cent a year, just under half the rate of human populations.[21]

As cropland expands, growing numbers of livestock press on a shrinking area of pasture. Soil is compacted. Vegetation cover degrades. Goats eat tree seedlings and stop forest regrowing. According to the Food and Agriculture Organization, in 1985 East Africa already had 2 per cent more livestock than it could support in the long term. North Africa's overstock was 11 per cent. The Sahel was carrying 52 million livestock units, 8 million more than it could sustain.[22]

Wood needs add to the pressures. Where woodland is still ample, there is no shortage of fuel and timber. But as forests are cut back to make room for farmland, the wood supply dwindles. Wood needs increase in line with populations. But at the same time, the area they have to be met from is shrinking. When the cut is more than the regrowth, the wooded area degenerates at an accelerating rate. Often the pressure is strongest not on forests, but on trees and copses scattered around farmland.

The needs for food, fodder and fuel all act together to degrade farm and woodland. Populations expand and cropland grows. Grazing and woodcutting exert heavier pressure on dwindling woodland and pasture. More and more fodder has to be grown on cropland. Because of the deepening fuel shortage, potential fertilizers are burned as fuel instead of being restored to the soil. Dung and crop residues combined provide a quarter of India's energy needs and 58 per cent of Ethiopia's. In Bangladesh, cow dung and crop residues supply four fifths of rural energy.[23]

At the end of this descent, cropland is producing food, fodder and fuel from declining reserves of fertility, and former woodland is reduced to barren waste. Only technological change – inputs of fertilizer, conservation or agroforestry – can reverse the results of rapid population growth.

A complex of causes

Technology is just as important as population growth in speeding land degradation. Ploughing up and down hill provides runways for rainfall and causes far more erosion than ploughing across the contour. Sparse planting and poor fertilization thin plant cover. The building of deep tubewells in pastoral areas concentrates dry season herds in a small area. The use of heavy machinery can severely damage laterite soils. Overwatering and poor drainage are the main causes of salinization.

By contrast, there are forms of terracing, mulching and intercropping that can cut erosion to levels almost as low as if the field were left wild.

Many other factors affect the choice of technology. Forms of ownership have a strong influence. Any form of insecure tenure

– whether a one-year sharecropping tenancy or uncertainties on state-owned land – works against soil and tree conservation.

Prices have an impact too. Where farmers are poorly paid for their crops, the value of the land that produces those crops will be low, and little effort or cash will be spent in conserving the land. Low prices, in turn, may be due to Western protectionism and dumping of surpluses, which has lowered world market prices of crops. They may be due to artificial encouragement of cash-crop schemes by development projects, causing an over-supply. Or they may be due to government price controls, inefficient state-owned marketing monopolies, or overvalued exchange rates.

Adapting to increased pressures

Technology and population growth are not independent factors. We know that in most places – though not all – rising population density has forced changes in techniques of cultivation and fertilization (see pp. 31–2). Does it also bring out new techniques of soil conservation?

The evidence suggests that it does, eventually, in most places. But the process is slower. It does not happen quickly enough to prevent heavy soil losses in places with heavy rain showers or erodible soils. And there are some places where it does not happen at all.

When population density is low, fallow periods are long, and erosion is minimal. Even if twenty years of soil creation are lost in a single year, there are twenty years of fallow in which it can be re-established. Soil conservation measures are unnecessary.

As fallow periods decline, erosion increases. But it often goes unnoticed until a good deal of soil depth and quality has been lost. This is the phase of soil mining. Eventually the drop in yields is serious enough to demand attention.

At this crisis stage a response must be made. In fertile areas, where land is valuable, the response will usually be to conserve the land, though not before a good deal of topsoil or fertility has been lost.

In marginal areas, where land values are low, the response may be perverse. Since conservation can involve additional effort and expense, farmers may just abandon the plot and clear land somewhere else. They may migrate to a town or city. If they have nowhere else to go – or if they are offered a technology that boost yields as well as conserving soil – they will put effort into conserving marginal soils.

The adjustment mechanism does not operate smoothly or reliably. If it did, there would no serious soil erosion anywhere in the world. In fact there is erosion almost everywhere. Land degradation is the norm. Population growth accelerates it. Compensatory advances in conservation are introduced only after the damage has been done.

10

QUINTESSENCE OF DUST:
Kalsaka, Burkina Faso

The monotonous Yatenga plateau, hard under the ribs of the Sahara, desertified at horrifying speed. The first time I visited this part of northern Burkina Faso, in 1976, there were no more than a few scattered patches of desertification. In 1988, on a dusty 180-mile drive north of the capital, Ouagadougou, I could not find a single healthy landscape.

Former fields and grassy fallow had been reduced to bare crusted wasteland scattered with a pavement of mauve pebbles. It was as if the soil had leprosy. The bushes were dry and brittle, and the dead trees held out their lopped branches, pleading for moisture from the sky.

West of the sprawling regional capital of Ouahigouya, a ridge of magenta hills, stubbled with dead shrubs, rises unexpectedly out of the plains. Below it lies the village of Kalsaka, a scatter of tawny-coloured compounds around the marketplace with its straw awnings perched on forked tree limbs.

Like so many other Yatenga villages, Kalsaka has lost half its cropland to desertification. Twenty years ago the centre was wooded savannah where people gathered fuel and pastured livestock. Today it is a square mile of bare rock where not a single blade of grass will grow. And it is spreading. From its fringes gullies eat back into cultivated fields, leaving trees balancing on their bared roots.

To the west is the Sawadogo compound, seventeen small round huts capped with cones of straw, almost a hamlet in itself. Each man has his own hut, each wife shares one with her children. Head of the family Jean-Marie Sawadogo, fifty-nine years old, with little eyes and long chin, strides around in long blue kaftan and a woolly hat pulled down to his eyebrows. The daytime heat would broil a European brain. But for locals, who

are accustomed to summer temperatures of 40°C, the 30° of
January feels quite chilly.

In my father's time . . .

In his twenties Jean-Marie served in the French army. He
fought in Indo-China in the 1950s. Now he remembers it as a
green paradise.

He has green memories of Kalsaka too, forty and fifty years
ago. Everyone, in every age, believes the world was better in
their youth. But talk to almost any older person in the Sahel
and you will hear a similar tale.

'In my father's time we never had a bad year,' he told me.

We used to have tremendous rains. It would start in the
morning and rain until the evening, and still be raining long
into the night. Millet filled all the granaries and was piled up
outside the compound under the straw shelters.

When we were boys the forest was all around us. It was too
thick to penetrate or to cultivate. The wild animals were too
many to count. There were antelope, elephants, buffaloes. If
we went hunting, we would see too many animals to kill them
all, and kill too many to heap on one pile.

When I was circumcised the land was fertile. You could
farm the same piece of land for five years, with no fertilizer,
before resting it. There was enough space for you to leave it
for ten or twenty years before you came back to 'it again.

Gradually more and more of the forest was cleared around
the compounds. Then each clearing joined up with the next
and created the great openness you see now. Now we have to
farm the same fields year after year. A piece of land that used
to fill two granaries would not even fill one now. Last harvest
it would not even fill half a granary.

The soil has been carried away. When I was young, you
could dig a hole as deep as your body before you reached
hard rock. Now in many of the fields you reach rock if you
just dig as deep as your hand. We used to find good clay to
make bricks right next to the compound, but these days we
have to go two kilometres.

Year by year the rain has got less. Today when it starts, it rains for twenty or thirty minutes and then it stops. And it seems to be getting harder. The drops hit the soil harder. Sometimes we see it rain behind the mountains, but it doesn't rain here. Every year we worry about whether the rain will fall or won't fall. Every year we say, 'Last year we had more rain than this.' We used to grow rice, sweet potatoes, but now it's too dry for them. Fig trees won't grow here any more. The last kapok tree fell down twenty years ago. Today the hills are bare. The only animals we see are hares.

Population growth has been the driving force behind land degradation here. Jean-Marie's grandfather had seven sons.* His father, just one of those seven, had five. More and more land had to be cleared. The fallow period used to restore fertility. But gradually the fallow was cut from fifteen years to one or two.

No fertilizer was applied to compensate. For much of the year the cattle herds were farmed out to nomads, who took them north. The desert fringes got their manure. The inherent fertility of Kalsaka's soil declined. Crops were thinner, so they protected the soil less against wind and rain. Topsoil washed away more easily, reducing yields even further. The grass cover in the fallow period was thinner – yet there were more sheep and goats grazing on it.

The land shows the scars. A hundred miles south there are one or two patches left like it should be here at this time of year, waist high in golden grasses, scattered with shrubs and trees, the grass forming a solid shield against the battering rain.

The process of decay follows a clear sequence, like the progress of a disease. On the impoverished soil the clusters of grass grow more widely spaced. Cattle and sheep chomp the grass down to the stump, leaving small bare patches between the clumps. Goats eat bush and tree seedlings and strip saplings of leaves, before they can mature or set seed.

Rain and wind work on the bare patches, carrying away more soil. At this point the structure of the Yatenga soil retaliates with a terrible and unpredictable vengeance. Silt and clay

* Only males inherit land.

particles lodge in the crevices of sandy ones to form an impervious crust. Rain can no longer seep down, but runs off in sheets. Now the plants are not just poorly fed with nutrients: they are starved of water, even when it is raining.

The grasses die first, surviving only in slight hollows that collect water and soil from the surroundings. A year or two later the bushes wither and dry to rattling bundles of sticks. Trees hang on till last, fed by their deeper roots. But as the water table drops, they too give up the ghost. The dead patches infect the living: rain rushes off them in torrents, streaming across cropped fields, scouring topsoil, gouging out gullies.

The processes have been speeded by climatic change. The Sahel has been getting dryer for more than twenty years. The causes are complex – but deforestation is probably one of them. Rain that falls to the south used to evaporate from forests, and drift further north to fall again. Now much of it runs off into rivers, back to the sea.

The combined impact of reduced rainfall and crusting is devastating. Two drought years, 1983 and 1984, followed in quick succession. The next two years saw good rains and fine crops. Jean-Marie's millet heads were like thick truncheons, eighteen inches long and two inches wide. His two granaries were full to the brim, and the crop lasted all year till the next harvest.

Then, in 1987, the first rains were five weeks late. The men dragged out the donkeys and ploughed. The women sowed. Everyone waited for further rain. But it did not come again for twenty days. The germinated seed shrivelled in the ground. After the second rain they sowed again. Again it shrivelled. They sowed four times in all. It was the end of July before the crop sprouted, two months late. In September, before the grain could ripen, the rain stopped again. The millet ears were little bigger than pencils. The sorghum, which should have been heavy with round white grains, had empty husks, fit only for donkey feed. Jean-Marie and his wife, Henriette, harvested just one cartful of millet, enough for two months' food.

Death valley

As dense fallow thickets were cut back for farming and crusting spread, firewood, too, got harder to find.

I went out one morning with Jean-Marie's wife, Henriette,
and the other women of the compound, to find out just how
hard. We set out at 7.50 a.m. The women had been up for
almost two hours, and had already made two trips to the well to
get water. The air was chill. The Harmattan wind, wafting
south from the Sahara, veiled the sun with a fine chiffon of dust.
Wrapped in shawls, they set out in single file, twelve of them,
machetes and hand axes resting on coiled ropes on their heads.
Young girls went first, so they couldn't trail behind and get lost.
Three women had babies on their backs, tied in their skirt-
wraps. Bouncing at each step, they kept slipping and had to be
constantly retied.

They passed up beyond the ploughed fields, into a valley with
copses where they used to cut wood. But in 1984, alarmed by
the pace of desertification, the village council forbade the
cutting of live branches. So they walked on, up on to a
desiccated plateau between two lines of hills. After an hour the
valley narrowed, and they stopped in a ravine of sparse dead
trees between dry red slopes of stony scree.

Henriette cut crimson strips from a tree they call *kundrenyango*
– the bark is chewed for toothache and septic abscess. She
collected the leaves of a shrub used to treat diarrhoea. 'Many of
the herbs we used don't grow any more, and the healers can't
get their medicines,' she complained. 'Twenty years ago the
trees here were so thick you couldn't get through. There were
hyenas, jackals, antelopes. Now all you see is lizards and
scorpions.'

Collecting firewood is not just a matter of gathering dry
sticks. It is more like a military operation, scouting out targets,
attacking with an armoury of weapons and tactics. One mother
hacked a dead branch with a blunt axe. At each blow her baby
jerked. It took ten minutes to cut a wedge half way through.
Then she tore the rest off.

Henriette picked up a hefty rock and hurled it at a stump,
until its rotted roots broke out of the ground. One girl climbed
into a tree and macheted away, one foot wedged in a fork,
another bent back against the trunk. The limb gave suddenly to
a hefty push, but she kept her balance.

The younger women, completely fearless, scramble up the
sliding ravine walls of sharp flaky ironstone. Their feet slip on

sharp stones, they clutch at spiny bushes. One slithers down an eighty-degree slope and wrestles with a dead sapling, poised over a forty-foot drop. Some fall, roll down, dust the grit off their scratched arms and legs, and carry on. Accidents are common. Hands and feet get cut on every foray. Only ten days ago one woman, hammering a root with a boulder, missed and bashed her foot instead.

After two hours' hard labour, they assembled their haul in piles, thickest branches underneath, brushwood for kindling on top, and tied them together with ropes. The loads, twenty kilos and more, are too heavy to lift unaided, so they helped each other to hoist them on to their heads. Then they marched off home at the same brisk pace they came, never stopping once, though by now their mouths were as parched as the landscape, and the babies were crying for milk.

We got back at 11.30, almost four hours after we set out. The women make this same expedition two or three times a week.

Desertification killed the trees. The fuelwood needs of growing populations was removing even their corpses. In less than a year the valley would be pure desert, stripped even of the bones of the dead woodland.

Making every drop count

Water, too, has grown scarcer.

It's not just the decline in rainfall. The crusting of soil and the death of vegetation has had an even greater effect. Rain no longer soaks in, but rushes off down gullies and streams. The water table has dropped. Wells have to be continually deepened, and dry up earlier in the year.

Henriette is lucky. She can get clean water close to home for nine months of the year. But between March and July the well dries up. She has to walk over two miles and queue. The round trip takes three hours, twice a day.

Many nearby villages are worse off. At Guenda, four and a half miles from Kalsaka down a sandy track, the getting and conserving of water has become a hardship of a degree few people experience outside wartime.

The Ouedraogo compound houses nine related families. In

many parts of the world it would pass for a small village. The children sit around in the narrow alleys, quiet with thirst and hunger, coated with a second skin of dust.

'Water is the worst problem in our lives,' complains Kiemdé Ouedraogo. At fifty her face is still handsome, but her breasts are shrunken and collapsed from repeated nursing. Five of her ten children died in infancy, mostly from water-related diseases.

The people of Guenda have not suffered passively. 'Nine times our men have dug wells,' says Kiemdé. 'But nine times the earth fell in before they reached water.'

They need a cement-lined tubewell, but they can't afford the materials and skilled labour. Eight years ago they built a dam 200 yards long, out of large stones. In the centre they hollowed out a long basin to hold water. This is where Kiemdé gets water from July to February. Twice a day she comes, with her fat round water jug. She ladles it full with half a calabash, sticks in a wad of leaves to prevent sloshing, and loads it on the rolled-up scarf on her head.

But her family pay a price for the convenience. Animals drink at the pool. Their hoof-prints are moulded in the shore mud, side by side with bare human footprints. The water is the filthiest brew I've seen – a dirty reddish brown, soupy with fine clay, live with tiny aquatic creatures. Kiemdé sieves it through a scarf before drinking. Left behind on the cloth is a sickening menagerie of little hoppers, crawling mites, flip-flopping larvae. Bacteria would still pass through to cause disease.

Guenda's families long for sweet fresh water, free of dirt and bugs. But the dirty pond water is only half a mile from home, and their time is overloaded.

'We know this water is giving us guinea worm,' says Kiemdé. 'But we have no choice but to drink it.'

By March the dam pond has dried to sticky mud. From then until July, the women have to trudge nine miles a day, to Kalsaka and back, to get water from the same deep well that Henriette uses at that time of year. The round trip takes four to five hours daily, for a single pot.

'To avoid queues we get up long before first cock crow,' Kiemdé explains.

All night long my heart is heavy with the lack of water. If I sleep at all, I don't sleep well, because I'm afraid of not waking in time. If the cock crows, I have to jump up, because I know I'm already late. We take the small babies who are still at the breast with us. The others we leave at home crying for us, with the old women who are too old to walk.

We never go alone. We're always afraid when we pass a man. You never know what they might do. We get very frightened by the brick pits near Kalsaka. They say there are ghosts there, though I've never seen one.

Often we set off after dinner and walk by starlight, to avoid the day's heat. We sleep at the well, on our shawls, and draw the water at dawn, as soon as the wells are unlocked. They're kept locked all night, to stop people taking too much water.

The jugs are heavy when they're full. While we're walking our neck hurts, our heart beats fast all the way. Even when we're not walking, our shoulders, the small of our back, our breast bone, all hurt.

It is an arduous journey. They walk with such poised grace, but they often drop the pots if they're tired, or trip over stones, or skewer their bare feet on millet stubble. Then they go back and get another pot and start all over again. Like oxen chafed where the yoke rubs, the women carry the scars of their labour. Their leather-hard soles are ringed with long vertical splits. There is not one of them who has not been stung repeatedly by scorpions. Kiemdé once spent three weeks on her sick bed with a bite.

Water so painfully won has to be husbanded as carefully as rations on a lifeboat lost at sea. From March to June a single large earthenware potful has to last Kiemdé's family of four and their chickens and goats for the whole day. It works out at six pints each, for all purposes. Yet these are the hottest months, when the body loses most moisture through sweat.

'No one can drink enough to satisfy their thirst,' Kiemdé explains.

I give them one cupful, then they have to go away again until their thirst pains them. Only then will I give them another cup to calm their thirst. We eat only once a day in those months. I use one calabash of water for the millet porridge and another cupful for the sauce.

We wash our plates and hands before the meal. This takes about three cupfuls. But we don't throw this water away. We keep it for one of us to bathe in. We take turns. Today my husband will wash. Tomorrow it will be my turn, after that my older son, then the younger.

I asked her what they do with the waste water from washing. She laughed.

There isn't any. It's hardly enough to wet every part of your body. If there is any left over in the jug, we put it out for the chickens and the little goats about to die of thirst.

The obstacles against intensification

Kalsaka is clear proof that technology does not adjust smoothly as population growth presses on resources. There were delays and blockages. During those delays serious problems developed and wrought widespread damage.

As in Madagascar, we have to ask: why didn't people intensify their farming methods? Why didn't they use more fertilizer? Why didn't they invent ways of stopping the crusting? Certainly not out of hidebound tradition: African farmers adopt new varieties as soon as they can get hold of the seed, if they see it performing well on their neighbours' fields.

Just as in Madagascar, the main reason why farmers did not begin to use fertilizer was that there was no land shortage. Shifting cultivation was well adapted to the environment. The fallow period restored fertility, and conserved the soil as well.

Land shortages were developing as populations grew. But after 1973, as desertification spread, the transition to scarcity accelerated.

Scarcity should stimulate innovation. Yet the people of the area were caught in a technology gap. The initial problem was the absence of fertilizer to compensate for declining fallow periods. In the Sahel and other dry areas, this is a technical problem, not just a matter of lack of cash. In dry years using conventional fertilizer will actually lose money. And farmers at subsistence level cannot afford to risk a loss.

No high-yielding modern varieties were available for millet
and sorghum, the two main food crops of the area. Researchers
had developed seed varieties which outperformed local varieties
on decent soils with sufficient water. But they hadn't found a
single strain that would do as well as local varieties in a dry
year. And farmers had to be sure of surviving every year – not
just in wet years.

Why didn't they conserve their soils as the desert spread
around them? For the same reason that the West didn't do
more about chlorofluorocarbons until the ozone hole appeared.
At first they didn't realize what was going on. They didn't
foresee the consequences of their own actions. They weren't
aware of the speed with which local soils would crust until it
actually happened. Even when it did happen, they didn't
understand the causes. Many thought God was punishing them
for their wrongdoings. And when they did grasp what was
happening, they had no solutions.

'Things happened very slowly and we didn't notice them at
first,' Jean-Marie explained.

At the beginning of an illness, you don't realize it can do you
harm. It's only when you can no longer walk that you realize
you are really sick. When we saw that the land was dying, we
knew we had to do something. But we didn't know what to
do.

In conservation, as well as in fertilization techniques, they
were caught in a technology gap. Western technology tried to
prevent the problem, but failed. In the early 1960s a massive
European Development Fund programme used bulldozers to
build low earthen dykes along the contours. The hope was that
these would slow erosion by blocking the run-off of rain. But
this technology was not tested on local farms. The people were
not consulted. They weren't even told what was going on. The
bulldozers trundled through their lands like alien spaceships. In
practice, the dykes dried up the fields below them. Farmers dug
holes through them to let the water through.

They tried their own methods. The Yatenga is littered with
lumpy, heavy ironstone rocks. Farmers used to clear these from
fields and pile them along the boundaries. After a while they

noticed that the stones slowed the pace of water running off the land, and began piling them across fields vulnerable to erosion. This did some some good, but it had a flaw: the lines did not follow the contours of the land. Water built up at low spots and gushed through, carving gullies at the spill-over point.

The farmers needed some way of tracing the contours. But the Yatenga is flat. Slopes are so gradual that often the unaided eye cannot even tell which direction they fall in. It was not till 1983 that a local project, run by the British charity Oxfam, found a solution: a cheap water level made of a ten-metre length of clear plastic hose. When level points have been marked with stakes, the contour lines are drawn across the slopes. Then stones are piled along these lines, plastered uphill with mud. The rain, dammed back by the lines, has time to soak in, yet enough passes through the gaps in the stones to water the field below.[1]

Jean-Marie Sawadogo was the first man in Kalsaka to start building the lines. He began with his second largest field, on a slope that descends to a wooded valley. Six acres had lain barren here for a decade. He had nothing to lose except his labour. 'Everyone laughed at me at first,' Jean-Marie remembers. 'They said it was useless, I was wasting my time.'

As soon as they finished the first stone line at the top of the hill, soil began to collect behind the lines. Dead leaves and seeds lodged and sprouted. At the next rainy season the effect spread uphill, and soon the field was green. The following year they sowed it with millet. The yield was double that of his neighbours' without lines.

Erosion was halted. The task of reversing desertification was begun. Even the water level in the well began to rise again.

'They stopped laughing then, and asked how to build the lines on their own fields,' Jean-Marie grins.

The continuing pressure of population

The stone lines are a form of intensification. They conserve water as well as soil, and boost yields because more water goes to feed crop roots instead of washing down gullies.

Two prerequisites were essential before they could be

adopted. The first was land scarcity: a fairly high density of population on cultivable land. If there had been plenty of good new land to open up, people would have done so and abandoned the old land. The second was the discovery of a technology that was effective and cheap and worked under peasants' circumstances.

The Yatenga is still caught behind the other technology hurdle: the soil fertility problem. The lines boost crop yields, but precisely because of that they also boost the rate at which nutrients are sapped from the soil.

Jean-Marie, adaptable as always, has begun to try out ideas picked up from other sources. He has built an enclosure in which he is making compost. He waters it faithfully every morning with the remains of his washing water. But most plant waste is needed for goat and sheep fodder. The rest makes enough compost for only a small fraction of his land. He has also begun to plant trees along the edges of his fields, leguminous trees that fix nitrogen in their roots.

But every solution leads to a new problem. The goats roam everywhere. Last year they ate eight of the ten tree seedlings he planted. He now needs a solution to the goat problem. I suggested that he keep them tethered and fed them on cut fodder. But where would they grow the fodder? And who would do the cutting and carrying? He is still thinking about that one. The problem of adaptation has no end.

The prospects for slower growth

Population and land resources have not adjusted to each other in the Yatenga. Indeed, Burkina Faso's birth rate actually increased, from 46 per 1000 fertile women in 1965 to 49 twenty years later.

The central plateau – heartland of the once powerful Mossi empire – is twice as densely populated as the national average, three times more than the moister south-west of the country. In relation to its capacity to produce food with existing technologies, the Yatenga is already seriously overpopulated. In normal years local production meets only 40 per cent of needs. In the drought years of 1983 and 1984 it met only 21 per cent.

Yet the population is still growing at 2·8 per cent a year.
Burkina Faso's population is projected to rise from 9 million in
1990 to 23 million by 2020. The World Bank suggests that it
may reach as many as 52 million before levelling off.

The people of Kalsaka show no interest in family planning.
They won't even talk about it: every question I asked on the
subject was met by sheepish grins and refusals. The dispensary
is supposed to provide advice. Yet in Idrissa Ouedraogo's three
years as senior midwife, only three women and one man have
asked about it. Not one has followed her advice to use modern
contraceptives. Even if they did, she has no supplies. The
nearest source is fifty-six kilometres away at Séguénéga. 'They
think it's shameful to talk about it,' she says. 'If women ask me,
I tell them to send their husbands along as well. They say: no,
he won't talk, he will chase me away.'

If supplies were available, they might be used by some:
widows, women abandoned by their husbands, women who had
had their six or eight children and could be confident that at
least four would survive.

But these would make up only a small minority. Despite the
fact that they are pressing up against the walls of local
resources, they see no need to have fewer children. There are
always safety valves: migration to the south-west, or to Côte
d'Ivoire. Indeed, having children to migrate elsewhere and send
back some wages helps to eke things out in the Sahel. The more
you can manage to rear to a productive age, the more can
migrate, the more money and gifts you will be sent to help
support you.

Wider improvements in education, health and women's
status would be needed to induce more people to want fewer
children. And despite brave attempts under the Sankara
government, these improvements are still lacking.

Burkina Faso's female literacy rate of 6 per cent is the lowest
in the world, while the male rate of 23 per cent is second lowest.
Only one girl in five is enrolled in primary school, and even
among boys only one in three. One boy in every twelve goes to
secondary school, one girl in twenty-five.

Places at Kalsaka's primary school are in such short supply
that it only takes in pupils every other year. Jean-Marie's
seventeen-year-old daughter Antoinette could not enrol: she was

born in the wrong year. Girls are kept at home to help mothers
grossly overburdened with chores. For every girl at the school,
there are two boys.

Conditions for those who do get to school are poor. Classes
are huge, ranging from forty-nine children up to eighty-five.
Teaching methods are strict. Individual attention is severely
limited. All teaching is in French – a foreign language for Mossi
children. There is no budget to pay for anything except
teachers' salaries – nothing for teaching aids, posters, reading
books, games. Teachers have to buy their own chalk. Families
buy exercise books and slates. The basic French text book
comes at £4.50 or $9 – equivalent, in relation to the average
income, to $770 for a US family or £250 in Britain. Half the
children in Class One do not have one.

Classes get smaller as they get older. Out of every ten pupils
who start the first year, only six finish the full six-year course.
The nearest secondary school is eighty-six kilometres away in
Ouahigouya. The previous year only two Kalsaka children out
of forty-nine school leavers won places there.

Child mortality has to drop below one in ten before couples
can feel confident that their children will survive. In Burkina
Faso as a whole, one child in five dies before its fifth birthday.
In poor rural areas like Kalsaka the rate may be one in four, or
more. Health care is appalling. There is one doctor for every
57,220 Burkinabé, the second lowest rate in the world after
Ethiopia.

In 1985 the Burkina government began a programme to
bring primary health care to village level using barefoot doctors.
Some 7500 traditional midwives and the same number of village
health workers were trained to deliver preventive and basic
curative care. But the crash training course lasted less than a
month, and the basic medical kits were not topped up regularly.
The village workers are unpaid volunteers. Twenty-two of the
twenty-seven reporting to Kalsaka dispensary are illiterate.
They have been taught to keep records with visual symbols.
Many keep no records. The village worker for Guenda spends
most of his time away, prospecting for gold.

The nurse in charge at Kalsaka, Salam Sawadogo, showed
me his sparse cupboard. It should contain the essential
medicines to treat all but the most serious complaints of some

25,000 people. There was not a single antibiotic pill, not a single aspirin, no cough medicines or antacids. For wounds, there was one pack of lint. The last fifteen inches of bandage was used while I was there, on a youth who cut his shin with an axe.

Deliveries are supposed to arrive every three months. Salam says they have been coming six-monthly or less. The fridge, without which vaccines would spoil, gets enough kerosene to keep going for only twenty-five days of the month. Salam covers the cost of the extra fuel by selling empty vaccine phials.

If a drug or dressing is not in stock, patients have to go the village pharmacy and buy it. Many can't afford what they need. When more expert treatment is needed, they have to pay the full cost of transport to the medical centre at Séguénéga, thirty five kilometres away. At Séguénéga there is just one ambulance. If it wasn't busy, it could take them to the nearest hospital, fifty-one kilometres further on at Ouahigouya. They would have to pay for the petrol.

I asked Salam what happened if patients couldn't afford to pay the costs of treatment or transport.
'They go away and ask their family to raise the money.'
'And if the family can't pay?'
'They stay at home.'
'And die?'
'Sometimes.'

He described some recent cases. A woman who had tetanus after an abortion couldn't afford the course of injections. She died. A man needing an urgent operation for intestinal occlusion couldn't raise money for the transport costs. He went home to die. A pregnant woman, in need of a Caesarian because her pelvis was too small for the baby's head, was forced to have the child at home. It was stillborn.

Even those who can afford transport face a drive that is a bumpy three-hour dustbowl in the dry season, a six-hour mudbath in the rains. In the past two years four pregnant women had been taken for Caesarians all the 86 kilometres to Ouahigouya. Two lost their babies anyway. In the other two cases both mother and child died.

Most poor people leave most conditions untreated, in the hope that they will cure themselves. If that does not happen they consult a traditional healer. Only if all else fails and a life is at stake will they use the modern health care system. By then most patients are beyond the stage of cheap or rapid cure. Then heavy expenses can be made with little hope of a positive outcome.

Two-year-old Aminata Ouedraogo, granddaughter of Kiemdé from waterless Guenda, was weaned at fourteen months when her mother got pregnant again. From then on she drank polluted pond water. Over the next ten months she suffered four bouts of diarrhoea, each at least a week long, during which she ate little or nothing.

When I saw her she was painfully thin. She had a rasping cough and pained breathing suggestive of croup – fatal if not treated promptly. Four times Kiemdé carried her the four miles to the health centre. She had spent £14 on injections – equivalent, in relation to the average income, to $2400 for a US family or £780 for a British family. So far there was no sign of recovery.

11

THE INTERIM IS MINE:
Abidjan, Côte d'Ivoire

It would be impossible for the people of the Yatenga to survive on farming alone. Every Kalsaka family has other ways of earning money to buy the food their impoverished land will not produce.

The most recent is panning for gold. Gold fever gripped Burkina Faso since it was discovered there in 1985. But few are growing rich. Most of those who hit paydirt remain at bare survival level. It's not hunger for gold that drives them. Just plain hunger.

In Kalsaká, ore was first found by a boy herding goats in the hills. He noticed flakes glinting in the bed of a rivulet. That valley has now been transformed into an opencast mine. Deep gashes scar the outcrops. The lower slopes are pocked like a shell-torn battlefield. Older men, women and children scrape and sift loose earth in shallow holes. Younger men armed only with picks and shovels sweat down pits up to sixty feet deep.

There are no fenced-off claims. The land is open to all comers. A hole belongs to him who dug it, for as long as he can defend it. The better holes are worked in shifts day and night.

With nine of his friends Buréma, twenty-five-year-old son of Jean-Marie's nephew, Bukaré, picks and hammers the lilac, sulphur and alabaster rocks down a fifteen-foot pit. His eyelashes are mascara'd with dust.

Bukaré, aged forty-eight, complains that he's too old to dig. 'I haven't the strength for it. My son digs. He can give me money if he likes, but he has his own family to feed. He gives me rocks, but usually only the worst ones.'

Every day Bukaré lugs home a sackful on his shoulder. Every night until eleven, he sits in his unravelling grey sweater, crushing quartz by the flicker of a kerosene lamp. His mortar is the sawn-off end of a steel cylinder, his pestle part of the drive shaft from an abandoned car. He slams it down hard with a deep groan at each blow, continually sifting out the big lumps to pound again into gravel. Outside her own hut his wife Mariam, after a long day collecting wood and water, pounding grain, cooking meals, is at her grindstone grinding the gravel into fine powder.

A large 100-kilo sack takes two days to crush. Then Bukaré fetches water in an old green oil drum tied on the back of his bicycle. They spend another day washing off the earth until the heavier gold remains at the bottom of the bowl.

'If I get one gramme of gold from a sack of rocks, I can say I have done well,' Bukaré explains. 'Many times I do three days' work and get nothing at all.'

He held up his latest gleanings in a tiny glass phial, bought from the dispensary nurse. There was barely enough gold dust to cover the bottom. When he has enough for the head of a match, he goes to the gold dealers in Kalsaka marketplace. They give him £5 or $10 for a gramme – the product of three days' hard labour, not counting the days when nothing is found.

Mariam and Bukaré have other sidelines to make ends meet. One day I found Bukaré at his tiny wooden loom weaving cotton thread, spun by Mariam, into long strips six inches wide. These he would then sew into blankets. Each blanket takes two days to make, for which he will net £3 or $6. Another night I found him knitting woolly hats for sale.

All these schemes have the same function. Labour must supply what the land cannot. Bukaré is one generation younger than Jean-Marie. The process of subdivision by inheritance has left him with half as much land. But he has twice as many children still at home to feed.

'We never harvest enough to last all year,' he told me.

I consider it a good crop if we only have to buy millet for two months. This year we had to buy for eleven months. And so I have to work. It has to be work I can do at home. As head of the family I have to be around to see what goes on, to look after my family's health and security. If you have many in the family you've got to work till you sweat. If I didn't work all

day to feed them, my children would have to go out begging.
I couldn't stand for that.

Exports of men

Even working sixteen hours a day is not enough to feed all the
additional people that the area produces.

Gold is a recent export. Burkina Faso's longest-standing
export is its people. The French began the tradition. They used
men from the Yatenga to build the railway that curves down
from Ouagadougou to Abidjan on the Atlantic coast. Jean-
Marie's grandfather was conscripted as forced labour for that
enterprise.

What began as compulsion continued as commerce. French
and then African planters in Côte d'Ivoire, to the south,
recruited Mossi labourers to clear dense forest for cocoa and
coffee plantations. The Ivoriens grew rich, by African standards, by mining free rainforest capital with cheap Sahelian
labour.

Every young Mossi male would spend some years in the
south. It became an obligatory *rite de passage*, an initiation into
adulthood that every male above the age of sixteen had to go
through to avoid ridicule. Then they would return, buy a
bicycle and a radio, marry, and settle down.

Much migration is now permanent. It involves more than a
quarter of the country's annual population increase. Some
25,000 people every year move from one rural area to another
within Burkina Faso, mostly from the Yatenga to the less
densely populated and wetter south and west. Another 40,000
emigrate permanently, mainly to Côte d'Ivoire.

Migration siphons off surplus population – and provides
extra funds to those left back at home. But it also delays the
adjustment of population and technology to the resources in the
sending area.

'Many people have left Kalsaka,' says Jean-Marie.

But I won't leave. Our ancestors are buried around the big
baobab by my grandfather's compound. I wouldn't like to
leave them. But it's not just on their account that I stay on.

I'm the head of our family. I can't leave the young and the old to perish here. As long as it still rains a bit, I'll never leave.

In search of Ninsabla

Two of Jean-Marie's brothers have gone to Abidjan. Two of his sons are there. Many of his cousins and nephews have gone.

I asked about his brothers. He tore a sheet of squared paper from a school exercise book and wrote down the address of a friend through whom his brother Ninsabla could be contacted. It was a long shot. But as I was passing through Abidjan, I decided I would try to find him.

I was about to experience a little of the wonder, and the disorientation, of a Burkinabé migrant lost in the metropolis.

I left dry Ouagadougou overcast with the Harmattan dust, dust in my hair, dust in my camera, dust ingrained in my skin. Flying in over Abidjan, the green oil palm plantations had an unreal, almost hallucinatory brilliance. There was water, the endless water of the Atlantic, unbelievable after a place where the rivers only run in the rainy season. And the plateau, girdled in fast highways, with its towers and chic boutiques and antique shops, just like the adverts torn from French magazines that Kalsaka villagers use to brighten their walls.

The address was a poste restante in Abobo, the vast satellite city where Abidjan, confined by one of the few remnants of natural forest, bursts out again to the north. At the post office, the assistant looked up the friend's name in a yellowed ledger:

Ousseini Sawadogo. Care of SETAO

SETAO is a big construction company. I phoned the personnel department. There was no one in. I went in person. The head guard rang the office from the security gates. Ousseini was not a regular: he came in from time to time to ask if they had work, and they had no address for him. They hung up. Just as I was leaving the phone rang again. Someone had remembered that

Ousseini was working on a new building at Yopougon. I caught a cab.

Orange cranes overhung four white concrete towers of a huge new hospital. Another security guard in mirror sunglasses took me to the welding shop, then to the carpenters. No one knew him. He left me at the gate and searched the site.

Ten minutes passed, then fifteen. I'd already given up hope. But then the guard reappeared with a well-built man in white overalls, woolly hat, and pockmarked face. It was Ousseini. He had a nervous look, wondering what a European would want with him. He pointed to another crane in the distance. 'Ninsabla works over there, he'll be there till 5.30.'

With ten minutes to get there, we took off at speed. As we veered off the main road a police whistle shrilled behind us. The driver halted, waited. But the police car didn't move, so he drove on. The police leapt into hot pursuit and pulled him off the road. After an argument about misunderstanding, the policeman confiscated the taxi-drivers' licence. He would have to pay £20 ($40) to get it back.

We continued down the sandy track towards the crane, which hung over a SETAO repair depot. But the workers had just left in a van.

I fixed with the guard to come back after lunch the next day.

A precarious prosperity

Armed only with a common name – four out of five Mossi are called Sawadogo or Ouedraogo – I had rated the chances of finding Ninsabla at about one in twenty.

I had the advantage of literacy, without which the city's signs and streetnames are cryptic. I had money for telephones and cabs, without which I would have worn out a great deal of shoe leather. It might have taken an illiterate penniless Burkinabé a forlorn week or more to track him down. For that period he would have to sleep rough and dodge the police.

I could hardly believe it when a small, slim figure emerged from the machine shed, bright orange overalls and hands smeared with black grease, and offered a wrist to shake. He took me to the workers' rest-room and offered coke. He was emotional at meeting someone who had just come from the village he left thirty-five years before. He hadn't been back in the last ten years.

He told me his tale. Like most Mossi migrants, he'd started on a plantation, way back in 1954. After a few months, through a friend, he got a labouring job with SETAO, living in a makeshift dormitory on site. The managing director used him for a few errands and was impressed by his conscientiousness. He took him on as houseboy, then persuaded him to get apprenticed as a mechanic. In 1970 he married Awa, daughter of another Mossi migrant from the Yatenga. They now have four children. He earns the relatively handsome wage of 55,000 francs a month – £110 or $220.

He is still a diligent worker, anxious about taking time off repairing bulldozers to talk to me. So I fixed to see him that night.

From the bus terminal in Williamsville, he took me down a dirt alley into a steep-flanked valley. Sandwiched between blocks of modern flats and the motorway, in the shadow of the city's scrap-yards, is a dense clutter of shanties. Precariously poised towers of battered car bonnets and doors lean over the rust-roofed houses. We crossed a bridge of rotting planks over a stinking stream strewn with refuse and old car doors.

And there in all the squalor stood an incongruous bungalow, two bedrooms and a large living-room, with rendered walls painted greeny brown, smooth cement floors, hardwood doors and windows, and a new zinc roof. At the door Ninsabla posed, proudly and stiffly, for a photo.

His chequered housing history is typical of Abidjan. He and Ousseini, friends from Kalsaka, built their first wooden houses in another squatter area, Adjamé. But in 1973 the city authorities bulldozed their homes, to make way for modern flats with rents they couldn't afford. As squatters without title, they got no compensation, no alternative site to build on. For seven years they were forced to rent.

In 1980, like so many of the Ébrié tribe who farmed Abidjan when it was still forest, the owner of the valley divided it into plots for sale. Together with another friend from Kalsaka, Ninsabla and Ousseini bought three neighbouring lots at £200 ($400) each. Locals call the settlement 'Dialogue', from the friendly medley of nationalities found here. The French city planners call it 'Scrapdealers'.

Ninsabla built his first house on the steep slope right under the scrap mountains – a tiny wooden shack of two rooms with an outside kitchen and a chicken coop. Then he scrimped and saved and built a line of three rooms, in rendered cement block. Instead of moving in, he became a slum landlord, renting the rooms out at £12 or $24 each per month. His own family stayed on in the original hut, in worse conditions than their tenants.

The income was an insurance against the times when SETAO hasn't enough work and lays people off. But he saved most of it, and began to build a more luxurious house for his family. The bungalow took four years to build, at a total cost of £1200 or $2400.

In the midst of a chaotic and insalubrious slum Ninsabla and his friend Ousseini created a small island of order: a quadrangle of houses around a sandy open patch, with some of the original papaya and bananas still in place. In most Third World cities neighbourhoods are ethnically arranged. New arrivals move in near friends from back home. Ninsabla's courtyard is a miniature suburb of Kalsaka. He rents rooms to Jean-Marie's other brother Siddiqi, a taxi-driver; to Amadou Zalé, night-watchman; Adama Belem, brick maker; and Amadou Yampa, a trained carpenter working as a nightwatchman.

The improvements continue. Ninsabla has almost completed a cement block latrine, two showers and a well.

But his situation remains precarious. The division into lots was privately done. Ninsabla and Ousseini applied for official sanction from the city government. They were still waiting. If permission is not forthcoming; if the city needs the land for development, or decides the quarter is too chaotic or the river too insanitary, it could raze their houses to the ground for a second time without compensation.

For the moment he has hard won a standard of comfort and prosperity that Jean-Marie, depending on the charity of his migrant sons for much of his food, can only dream of.

Yet Ninsabla's heart has never left Kalsaka. 'We stay here because our kids go to school here, but we never forget back home. I miss Kalsaka. I often think of home. Sometimes I think about it so much that I can't eat or sleep.'

The sorted city

Abidjan's growth has been among the fastest in the world.

It was established, on the banks of the Ébrié lagoon, as the terminal for the Niger railway. In 1930 it had only 30,000 inhabitants. By 1963 it had grown to 290,000. Twenty-one years later it had rocketed to 1·7 million. Its average growth rate over fifty-four years was an astonishing 9 per cent. Two-thirds of that was due to migrants, many from the poor dry Sahel to the north or from economically stagnant Ghana and Guinea. A lot of its magnetism is due to the fact that for a long time investment was focused here: Abidjan has two-thirds of all modern sector jobs in Côte d'Ivoire.

The city has something of Ninsabla's tidiness, at least superficially. Everything has its special quarter. In the laundry area, acres of sheets and trousers spread out to dry on the motorway verges. In the scrap-metal quarter, pyramids of junked cars stretch across dozen's of dealers' yards. Even the tomato sellers in Abobo market range their wares, over thirty metres of pavement, in order of ripeness from green to blood red.

Residentially, too, it is sorted. Western-style houses, ranging from the ambassadorial mansions of Cocody to the basic modern flats of Williamsville, make up 30 per cent of the housing.

Much further out, squatting has been turned into a form of town planning. Here we find what the planners call 'evolutionary' housing, which accounts for more than half the stock. Tin-roofed buildings in breeze block and adobe are ranged along dirt roads. These are laid out in rectangular blocks, ready for the day when the city gets round to laying sewers and water pipes and tarmac. Many of the houses here are speculatively built, for renting to the newest waves of migrants. A lot are illegal – but even these are laid out in geometric networks, so they stand some chance of retrospective authorization.

The city's order has been achieved at a heavy price. Chaotic settlements that stood in the way were bulldozed. The residents were evicted, left to their own devices to find another niche. Often they were evicted again two or three or four times.

Some slums have survived the demolitions, though their days

are numbered: Vridi, next to the canal that links the lagoon to
the sea, first settled by the canal diggers; Washington, on a
steep hill near the motorway; Zimbabwe, the fish-smokers'
quarter; Zoe Bruno, named after the farmer who let his land out
to his friends and their friends: a chaos of hutments with mud
paths, but self-provided with its own wooden cinema, church
and school; High Tension, on vacant land under the electricity
pylons, a long thin wasteland of squalid housing, abandoned
cars, and low-skilled workshops.

Port Bouet, once a village of Ghanaian fishermen whose
painted canoes still line the beach, is now a motley of tribes and
nationalities. In the sandy alleyways Liberian women smoke
shark steaks, a family of Tuareg artisans from Mali make
sandals and leather-covered boxes, Guineans play the board
game ayo. The beach serves as playground for the children, and
as toilet for everyone. People squat on their haunches in full
view of passers-by. By noon it is thick with excrement and
paper, waiting for high tide to flush them away.

Provisional Lot 203

Large sections of Port Bouet have been demolished and the
occupants scattered to the four winds, some of them repeatedly.
Some moved to Marcory Poto-Poto. When I came to this area
in 1976, it was an insalubrious swamp backed by the tiny huts
of low-class prostitutes.

But then Marcory was drained and covered with sand. Many
Marcory refugees have taken up residence in temporary hamlets
on the drained site, for the brief interim before better housing is
built there and they will have to move again.

Ethiopia sprawls across the sand. The huts have tin roofs and
cement floors, but that is about all that can be said for them.

The rudimentary services are typical of Third World shanty
towns. Electricity snakes in, on an illegal pirate connection, to a
shaky 'pylon' carpentered from wooden joists and nails. From
here the individual wires spread out like the radials of a spider's
web. 'Subscribers' pay £2 ($4) per light bulb socket per month,
or £3 ($6) per plug. I met several whose radios had been
roasted by the supply.

Water is bought, from the few people who have taps, at an average of £12 a month for a few buckets a day. This is roughly what my family pays for unlimited water and sewerage in London. But there are no sewers in Ethiopia. The vacant areas serve as toilet. And as refuse dump.

Provisional Lot 203 faces the dump. The name sums up its own status and that of its inhabitants. There are seven rooms surrounding a courtyard, housing a radio repairer; a watch repairer; two nightguards; a tailor; a cloth merchant; a diviner; and all their families. They are all from Burkina Faso. They pay rent of £10 or $20 a month each to the landlord, another Burkinabé.

Tasséré Ouedraogo, his shelves lined with worn and dirty books in Arabic, casts fortunes with cowrie shells. He was first brought here as a forced labourer under the French, but escaped after a year. He was driven out of Port Bouet to Pro Domo. Evicted from there, he moved to a refuge for Port Bouet evacuees called Port Bouet II. Here he built himself a brick house. It was demolished. He had title and got compensation. But it was not enough to build another house, and he moved to Ethiopia. He works as a guard in Abobo, earning £60 a month. For that he spends four hours a day commuting and £13 in bus fares per month. He leaves home at four in the afternoon, and gets back home at eight the next morning.

Amadou Soré and his wife Aminata have been evicted three times. They now live in a single rented room in Provisional Lot 203, eight feet square. The roof leaks. The wardrobe is a row of nails on the rough plank walls. A double bed occupies almost half the space. The couple sleep here with their two year old child. A grubby sheet slung from a string provides makeshift privacy. Three other children sleep on blankets on the bumpy concrete floor.

The Sorés came down from a Yatenga village sixteen years ago. Amadou started as plantation labourer and woodcutter, then worked as a nightwatchman. When he got a job as a bulldozer driver he thought he had it made – but he broke his arm in an accident and lost the job. He became a nightwatchman again, but was sacked six years ago when recession began to bite.

Now he repairs watches in a tiny booth next to the house.

Aminata and her children spend all day by the road waiting for customers for her tiny stock of cigarettes. The couple are lucky if they make £60 or $120 a month between them. They spend £10 a month on rent, and £6 on charcoal. They pay another £5 for one light bulb and one two-pin plug for the radio. Water costs them £12, for three buckets a day. Food takes the rest of their income.

'We're like slaves here,' Amadou complains. 'The Ivoriens treat us Mossi worse than animals. The police always ask you for your identity card, then confiscate it and ask for 2000 francs to give it back [£4 or $8]. I would go home tomorrow. But I can't afford the fare.'

One or two million Burkinabé – no one knows the exact number – live in Côte d'Ivoire. That is equivalent to 10 or 20 per cent of the Burkina population, 8 to 17 per cent of Côte d'Ivoire's population. A quarter of Abidjan's population come from Burkina or neighbouring Mali.

So the poverty and population growth of the Sahel swell the booming cities of the coast.

12

THE QUICK OF THE ULCER:
the environmental impact of cities

These days the heads of families have crept into
town, abandoning sickle and plough. To feed us we
hire people to import grain from Africa and Sardinia,
and our wines come by ship from Cos and Chios.

Varro, second century BC[1]

Urban settlements are as old as agriculture. Big cities are a
more recent phenomenon.

In ancient and medieval times, even the biggest cities in the
Western world rarely numbered more than 40,000, less than
half the size of Tunbridge Wells. Athens, greatest trading city of
the Greek world, housed perhaps 300,000 people at its height in
431 BC – the size of Sacramento.[2]

The biggest city of the ancient world, Rome, drawing on the
produce of a vast empire, reached around 650,000 by AD 100.
The size brought familiar physical and social ills. In the second
century AD the satirist Juvenal complained of jerry-built
tenements, high rents, muggings, slop pails emptied out of
windows on to the heads of passers-by, and the terror of fires,
falling roof tiles, collapsing buildings. When Julius Caesar
banned vehicles from the streets in daytime, wagoners had their
traffic jams at night. The poet Martial wrote of urban
alienation: no one in Rome was so physically near, yet so
distant in human terms than Novius, the man in the window
across the alley.[3]

Rome was a grand exception, difficult to sustain in a society
based on agriculture and wood energy. Up until the end of the
first millennium of our era, Asia housed most of the world's
biggest cities. The life-span of these vast metropolises was
closely linked to the stability of the empires whose taxes and

tributes maintained them. Changan in China, in the eighth century AD, and Baghdad in the tenth, both briefly reached one million inhabitants, but soon declined. When the Spanish conquered Tenochtitlan, capital of the Aztec empire, it was bigger than any city in Spain.

Western cities were modest in comparison. Constantinople, at its height under Justinian, had a population of some 300,000. In the early fifteenth century only five European cities – Constantinople, Paris, Naples, Venice and Milan – numbered more than 100,000 inhabitants.[4]

Three main factors limit the size that cities can grow to. One is the number of jobs outside agriculture: in administration, armed forces, trade, services and manufacture. Another is the food surplus that agriculture is capable of producing, above the needs of its own workforce. The third is the availability of wood and fodder for building, transport and energy.

In the Orient and ancient West, cities were mainly imperial or administrative centres, subsisting on tribute or taxes from agricultural subjects. Since the surplus per agricultural worker was small, the size of imperial cities was limited by the size of their dominion.

In Western Europe, from the Middle Ages onwards, free cities began to grow outside the feudal framework. They became centres of manufacture and service, living on what they exchanged with rural areas. From the fourteenth to the seventeenth century there was a slow but steady growth in city size as trade and the productivity of agriculture increased.

Wood, not food, became the major constraint on size. Timber to build houses or ships was valuable enough for long-distance transport to be worth while. But vast quantities of wood would have been needed to fuel industry on a large scale. To lug this over hundreds of miles would have been uneconomical.

It was only the shift to a denser fuel – coal – that made the industrial revolution possible. And the industrial revolution made the urban boom possible. Labour was siphoned away from rural areas. Farmers were forced to increase output per worker by adopting labour-saving technology.

The modern city is – and remains – the child of fossil fuels. Coal enabled cities to make a quantum leap in size. The impact of the industrial revolution on world urbanization was dramatic.

Cities bigger than 200,000 numbered no more than a couple of dozen in 1800. By 1900 there were 148 of them. London numbered fewer than 200,000 inhabitants at the end of the seventeenth century – enormous for its day. But by 1800 it had reached 1 million. It more than doubled to 2·36 million in 1850, and by 1910 had doubled again to 4·5 million.

The urbanization of the world

Advances in transport and food preservation allowed cities to expand further. Canals, then railways and steamships made it possible to fetch grain from expanding hinterlands. Canning and refrigeration allowed meat and vegetables to be imported over long distances. The world became the city's market garden.

Urban railways spurred city growth along their routes. Dwelling and workplace no longer had to be close together. The internal combustion engine and the shift to oil allowed further urban expansion. The car freed the better-off from dependence on mass transit. Cities and suburbs sprawled in all directions.

Cities have gone on growing faster than the population as a whole. The data here, as in every other sphere, are often unsatisfactory. The United Nations Population Division accepts local definitions of an urban settlement, and this varies widely between countries. So the data can't safely be used to compare urbanization levels between two countries. But they can serve as a guide to trends over time.

Between 1950 and 1990 the world's towns and cities grew more than twice as fast as rural areas. Back in 1950 only twenty-nine people in every 100 lived in cities. All the urban areas in the world had only 734 million inhabitants, and only two cities – London and New York – housed more than 8 million people.[5]

The world of 1990 was almost unrecognizable. The world's urban population had more than trebled, to 2,390 million, and forty-five people in every 100 lived in towns or cities. There were now twenty hyper-cities with more than 8 million people. Fourteen of these were in the developing world, where in 1950 there had been none. Developing countries, which in 1950 had only 39 per cent of the world's urban population, by 1990 had 63 per cent.

The future is even more urban. Over the 1990s no less than 83 per cent of the world's population increase is expected to take place in urban areas. During the decade towns and cities will add 81 million people every year – equivalent to around ten extra cities the size of Moscow, Delhi, Paris or Lagos. While annual additions to total world population are expected to peak in the 1990s, urban increments will go on growing – reaching 95·5 million per year between AD 2020 and 2025. By then two people out of every three in the world will be living in urban areas.[6]

Hyper-urbanization

Some 85 per cent of the growth in world urban populations in the 1980s was in developing countries. Here city growth has been faster than in the West.

Europe's cities grew by only 2·5 per cent a year at the time of their fastest growth, between 1850 and 1920. Urban areas in developing countries have grown at annual rates of between 3·5 and 4·8 per cent a year since 1960. Africa has seen the fastest city growth: the annual growth rate has remained above 4·5 per cent since 1950.[7]

The Third World's swing from rural to urban has also been faster. Europe took seventy years to move from 15 per cent urban to 35 per cent. Developing countries completed the same shift in half that time. In Europe urbanization marched in close step with the the growth in manufacturing employment. In developing countries urbanization was initially far faster than the growth in manufacturing, with a much higher share of employment in services and administration.[8]

Much of the urban expansion has gone into the cancerous growth of primate cities. African urbanization was least balanced. In 1980, 36 per cent of Africa's urban population lived in the bigest city in each country – up from 28 per cent in 1960. In Latin America the largest city share rose from 27 per cent to 29 per cent over the same period. Mexico City received 40–49 per cent of all migrants within Mexico between 1940 and 1960.[9] Between 1964 and 1973 Bogotá received almost a half of all rural to urban migration in Colombia.[10]

The proportion of the urban population living in cities of more than half a million has shot up. In developing countries as a whole, it rose from 31 per cent in 1960 to 44 per cent twenty years later. In Africa it has exploded from 6 per cent to 41 per cent over these two decades. The global number of cities of this size grew from 111 to 278. In Africa their numbers multiplied nine fold.[11]

The fastest growth has been at two extremes of the urban scale: in towns with 100,000 to 250,000 people and in big cities with more than 4 million. In 1950, there were only ten cities with more than 4 million inhabitants. By 1990 there were forty-three – thirty-three of them in developing countries.[12]

Industry and services benefit when they cluster in big cities. They can share the same skilled workforce, banking facilities, transport and stores, and dealer networks. And they benefit from a large local market. These economies of clumping go directly to business balance sheets and workers' pockets.

But there are disadvantages: crime, congestion, pollution within the city and in its wider region. These *dis*-economies are suffered by consumers and residents within and without the city. These widely scattered losers from urban growth are less well organized to make their voice felt, or to make sure their costs are accounted for.

Environmental problems within cities

The city itself is the daily environment for almost half the human race. And in the Third World that environment has deteriorated as the numbers living in cities increased. Population growth has outpaced adequate housing, livelihoods, roads, transport, water, sanitation, waste disposal and health services.

Bad housing is the most visible problem. The typical Third World city is an island of concrete, steel and glass surrounded by an ocean of wood, tin and plastic. In 1980 around one in

four urban residents in Asia lived in shanty towns; two out of
five in Latin America and the Near East; two out of three in
Africa.[13]

In some cities the share is much higher. In Addis Ababa 85
per cent of people lived in squatter settlements in 1980, and 70
per cent in Luanda. In Ankara the proportion was 51 per cent,
in Manila 40 per cent. In Latin America the share ranged from
32 per cent in São Paulo to 59 per cent in Bogotá. Where the
percentage living in squatter settlements is low, it is more often
a sign of ruthless police clearance than of adequate housing
provision.[14]

Squatter housing is by definition temporary. Often, like the
houses built by the first two litle pigs of the tale, it is easily
blown down, leaky and draughty. This flimsiness is not always
due to low incomes. Often it is the result of insecurity of land
tenure. People who are liable to be evicted without compensation will not risk their savings in a solid house. Wherever
squatters are given security of tenure, they gradually upgrade
their housing as savings allow. In Lima's desert suburbs, mud
floors cede to cement, straw or plywood walls to brick or breeze
block.[15]

Overcrowding is endemic. Whereas in Western countries
there are, on average, only 0·66 persons per room in each
dwelling, in urban Mexico there are 2·2, in El Salvador 2·4. In
urban Sri Lanka and Pakistan the figure is 2·7 persons per
room. These figures average the pauper's hovel with the
millionaire's mansion. At the lower end overcrowding is far
worse. In India, the average is 2·8 persons per room. But two-
fifths of the urban population live in one-room shelters with an
average occupancy of five persons.[16]

In developed countries, higher incomes and slower population growth has permitted a steady increase in the number of
dwellings per person. In developing countries the number of
permanent dwellings has not kept pace with population growth.
In 1975–9, for every 100 new households, only 53 new
permanent dwelling units were added. A decade later provision
had fallen to only 38 per 100: the other 62 were having to make
do with shanty-type accommodation. In sub-Saharan Africa,
only 11 permanent dwellings were provided per 100 new
households in 1975–9.[17]

This shortfall in permanent housing is not just the fault of
population growth. Just as often, Western-modelled official
standards raise the cost of formal housing beyond what most
people can afford. Indeed the whole concept of 'housing
provision' is misleading in the Third World. Most people
provide their own housing, or rent from small-time landlords
one rung up the ladder.

High rents and fares, low wages

The growth of city populations changes the layout of cities. The
price of land in inner zones rises. Offices and larger shops take
over the centre. Most housing and industry shift outwards.

Homes and places of work grow further apart. Family
budgets are squeezed in a vice between rising rents near the
workplace – and rising fares from further out. In Abidjan 56 per
cent of jobs are on the south bank of the lagoon – but 66 per
cent of the population live on the north bank. North–South
fares average 6500 francs per month – 11 per.cent of a skilled
wage. In Mexico City the average share of wages spent on
transport rose from 9·4 per cent in 1968 to 13·4 per cent in 1978.
In some countries the poorest families spend up to 30 per cent
of their income on transport.[18]

As the city sprawls, the very poorest can't afford fares from
outlying areas and turn to illegal squatting closer to job
opportunities. Some live on pavements or under bridges. Others
knock up hovels where they can, liable to be moved on from one
day to the next. Vacant land within reach of the city centre is
usually dangerous, insanitary or polluted. In Ankara squatter
housing is built on unstable stone platforms piled against steep
slopes. In Rio, Bangkok and Manila shanties stiltwalk over
stagnant lagoons. In Bombay they cling to the terraces of
abandoned quarries and cluster round swamps.

Often the city economy does not generate enough jobs paying
adequate wages. Open unemployment is uncommon. Only in a
few countries (Sri Lanka, Jamaica, Barbados, Trinidad and
Tobago) does it reach 20 per cent or more. In India urban
unemployment rose inexorably, from 2·6 million in 1966, to 9·6
million ten years later, to 33 million in 1991. But in most

developing countries, social security systems exist only for public employees or big firms. Few others can afford to be unemployed.[19]

Hence the spread of small and low-tech enterprises and trades that bloom, and often wither, outside the sunshine of government favour. These informal businesses rarely benefit from government measures to encourage domestic industry. Indeed they are often persecuted and harassed. Yet frequently they flourish against all odds. They account for around a third of all urban jobs in Asia and Latin America, and over a half in Africa. In the debt-depressed eighties, employment in the formal sector rose by only 3·2 per cent a year in Latin America. Informal jobs expanded more than twice as fast.[20]

The incomes of employees in the informal sector are low. In five African cities, between 43 per cent and 100 per cent of unskilled workers earned less than the official minimum wage. In Colombia self-employed men in the informal sector earned 80 per cent of the average wage, and women only 50 per cent.[21]

Such low wages suggest an oversupply of labour, stoked by immigration from rural areas and the city's own population growth. But again it is not as simple as that. Governments often provide artificial support for large-scale enterprises and harass small and informal enterprises.

Urban water needs are still poorly provided for the majority. In 1988, two years from the end of the International Drinking Water Supply and Sanitation Decade, only 42 per cent of African town dwellers had piped water in their houses. Another 32 per cent had to get their water from communal standpipes. In urban Latin America 14 per cent used standpipes. The urban proportions without easy access to clean water ranged from only 2 per cent in the Eastern Mediterranean and 12 per cent in Latin America, to 26 per cent in Africa and 29 per cent in Asia.[22]

Sanitation is in far worse shape. While only 10 per cent of town residents in Latin America have no form of sanitation, the share in Africa was 46 per cent and in Asia 59 per cent. In Manila only 11 per cent of population are served by sewers. Road gutters, open ditches and canals serve the rest. In India only 209 towns and cities out of a total of 3119 had partial or full sewerage systems in the mid 1980s.[23]

The rich get the profits, the poor get the poisons

On a cloudless January afternoon, Bombay is encased in a slate-grey photochemical smog, as if sweltering inside its own private greenhouse. From a plane, only the surrounding hills are visible, islands in a lake of mist.

In Mexico City total emissions of local air pollutants rose by 45 per cent between 1972 and 1983. Street booths sold puffs of oxygen. Dried and powdered excrement wafted through the air.[24]

City air pollution became a severe problem when coal replaced wood as the main heating fuel. As early as 1661 John Evelyn complained of the effects of sea-coal: 'It is this horrid Smoake which obscures our Churches, and makes our Palaces look old, which fouls our Clothes.' He compared London to Lake Avernus, whose volcanic fumes suffocated any birds that flew over it. For centuries London was famed for smog – one fog, in November 1879, lasted till the following March.[25]

Since then urban air pollution has improved in most developed countries. London has not had one smog in the twenty years I have lived there. Sulphur dioxide emissions in London fell by 8 per cent a year over the twelve years to 1985. In New York they dropped by 7 per cent a year. But in many developing countries the situation has worsened. In Calcutta sulphur dioxide emissions rose by 5 per cent a year between 1973 and 1985, and in Hong Kong by 8 per cent. In Tehran they grew by 10 per cent every year, and in New Delhi by 20 per cent.[26]

As for smoke particles, seventeen of the worst twenty cities for the period 1980–4 were in developing countries. Seventeen of the cleanest twenty-one were in developed countries. Where New York had an average of 53–89 microgrammes of suspended particulate matter per cubic metre, and Frankfurt between 22 and 44, Jakarta, Delhi, Beijing, and Shenyang all had ranges above 200. Lahore had an average of almost 800 in 1981.[27]

In cities humans are the endangered species. Urban air pollution has serious effects on health. In Chinese cities the incidence of lung cancer is four to seven times higher in cities than in the nation as a whole. Three out of five people in Calcutta are believed to suffer from respiratory diseases related to air pollution. In Mexico City air pollution is a likely cause or contributor to 90 per cent of respiratory illnesses and infections.[28]

But the dangers do not weigh equally. The better-off buy into areas safely distanced from smoke-stacks: areas with piped water, sewerage connections, paved roads. They travel by car or taxi, from door to door. Their political clout ensures that when city budgets are limited, their needs are first met.

It is the poor who are in the firing line. It is the poor who live on dangerous hillsides and unhealthy swamps. It is the poor who have diarrhoea, the poor whose houses are swept away in landslides and earthquakes. It is the poor who live in the black shadow of factory chimneys and bathe by factory waste outlets.

It is the poor who live without piped water or sewerage connections. It is the poor who so often have to buy their water from vendors. The people with taps in their kitchens pay anything from four to a hundred times less. Some of them sell their subsidized water to the waterless, at a fat profit.[29]

It is the poor who live in areas without health centres to deal with the health problems that their environment creates. In Karachi, for example, there are five times as many people per dispensary in squatter areas as elsewhere. Non-squatter areas have four times as many hospital beds per person as squatter areas.[30]

Unable to live near work, forced to use slow public transport, it is the poor who suffer most from city congestion. In Bogotá, the average low-income adult spends 127 minutes per day travelling to and from work. High-income travellers spend 44 minutes less.[31]

And that congestion is caused mainly by cars, taxis or rickshaws used by the better-off. Dhaka's rush hour is a suffocating chamber where you come out desiccated and kippered. Beggars profit from the jams by importuning drivers. The poorest walk. Low-paid clerks strap-hang in cigarette-packed buses that bear the scars of hundreds of collisions. The better-off ride in rickshaws whose suicidal pedallers pass on the wrong side of the road, jamming all comers to gain a few seconds' advantage.

Cars take up nine times more road space per passenger than buses. In Karachi cars cover only 44 per cent of passenger miles but use up 83 per cent of road space. In Manila, cars and taxis account for 54 per cent of all vehicle trips but contribute only 30 per cent of person trips.[32]

With worse environmental health problems and poor health care, health in squatter areas lags far behind city averages. Among Manila squatters, infant mortality is three times higher than among non-squatters. Typhoid cases are more than four times more frequent. In Calcutta the death rate from tuberculosis is more than twice as high in squatter areas. Deaths from dysentery are three times more frequent.[33]

Environmental impacts outside cities

Cities are increasingly important shapers of the environment *outside* cities. Urbanization increases the environmental impact of growing populations on the natural environment.[34]

Pre-industrial rural areas provide the major bulk needs of life for themselves. Cities import most of their food, fuel, building materials and water. Provisioning cities costs in other resources. Food has to be packed, processed, preserved, transported. In agriculture, surplus production per worker must rise to meet the needs of urban dwellers. This can only be done by increasing the input of commercial energy into agriculture. In developing countries the energy used per worker rose from 51 kilos of oil equivalent in 1972 to 95 kilos in 1982. In North America each farm worker used a colossal 26 tonnes of oil equivalent in 1982, up by 7 tonnes in ten years.[35]

The city competes for resources with agriculture and forestry. Cities usually grow on and eat up prime agricultural land. Agriculture compensates by moving into more and more marginal areas (see pp. 107–9).

The city remains a major user of fuelwood and charcoal in many developing countries. But as forests close to major cities shrink, the wood supply-lines lengthen. To meet Delhi's firewood needs, no less than 12,423 railway wagons of firewood arrived at Tughlakabad railway siding in 1981–2, carrying 613 tonnes a day. Most came from Madhya Pradesh, 700 kilometres away.[36]

Above all, the city pumps out waste – solid, liquid and gas –

in far greater quantities per person, and in far higher densities, than rural areas. We shall look at these in detail in Chapters 13, 14 and 15.

Is national population growth responsible for city–population growth?

In theory cities can expand – or shrink – through migration alone, while the population of the country as a whole remains static. In practice the growth of urban and of national populations is strongly related. Cities grow faster when national populations grow faster. Over the five years from 1975–80, in 48 countries with the fastest national population growth, urban areas were growing at an average 6·1 per cent a year. In the 49 countries with slower national population growth, urban areas were growing at only 3·6 per cent a year. In statistical germs, the national rate of population growth 'explained' 45 per cent of the differences in urban growth over the 1975–80 period.[37]

These results are significant. They mean that city growth is not simply an isolated phenomenon, but is linked to the country's overall growth rate. This is not a surprising discovery. City birth rates are usually lower than those in rural areas. But in countries with high birth rates, city birth rates will also be relatively high. The extent to which the country was already urbanized is also a factor. Where the percentage of people living in cities is low, the rate of city growth in following periods is faster.

But the links go further than this. In countries with faster total population growth, city growth appears to have a tendency to race ahead of national growth by a bigger margin. In 1975–80, the urban share of total population grew at an average of 2·8 per cent in the 48 countries with faster population growth. In the slower growing countries the urban share increased by only 1·8 per cent.

Urban population growth is a combination of two elements: the city's own natural growth, and migration from other areas. At times migration has had a huge impact on rural and urban areas. The United Nations Population Division estimated, on the basis of census data from the 1960s and 1970s, that about a third of the rural population increase in Africa and Asia, and 58

per cent in Latin America, was lost to migration, or the
swallowing of rural settlements by urban. During this period
urban areas owed much of their growth to in-migration and
reclassification of rural areas: between 36 per cent (in Latin
America) and 47 per cent (in Asia).

More recently, natural population growth among people
already living in cities has made up the lion's share of city
growth. By 1980–5, natural population growth among people
already living in cities made up at least three-fifths of urban
growth in developing countries.[38]

But the contribution of population growth does not stop
there. For it also fuels migration.

Push or pull?

People leave rural areas and head for cities for one basic reason:
they believe that their own fortunes, and those of their children,
will improve if they do so.

The move is not lightly made. In some cases there is no
choice. Where land and jobs are scarce, there may be no other
way of surviving. Ninsabla Sawadogo went to Abidjan by
choice. Yet he did not have much option but to stay there to
raise a family. The people of Kalsaka depend for survival on
cash sent home by those who have gone away. If all the
migrants returned, everyone would starve. Much contemporary
migration from marginal areas is of this compulsory nature.

In cases where survival is not at stake, migration is a
calculated gamble. Migrants compare conditions at home with
what they know of conditions in the city. Both sides of the
comparison are important. The steeper the gradient, the faster a
ball rolls down hill. The greater the differentials in income and
other opportunities between urban and rural areas, the more
rapid is the migration.

The cities' pull is one side of the equation. Migrants often cite
better educational opportunities for children as a reason for
moving. In poorer countries most secondary schools are sited in
urban areas. Rural pupils often have to board. Even at primary
level the gaps are wide. More urban schools offer the complete
course, while rural classes are cut short.[39]

Health provision is better in urban areas. Out of sixty-three developing countries cited by World Bank economist Johannes Linn, only one – Tunisia – had more hospital beds per person in rural areas than urban. In the other sixty-two, the number of rural persons per bed was anything from 1·8 times the urban ratio for Ghana and Iran, right up to 74 times for Botswana and 143 times for Libya. Even primary health care systems, designed to improve service to rural areas, have their better-equipped, better-trained staff in urban centres.[40]

The most important determinants of health are clean water and adequate sanitation. Here, too, urban areas are favoured. As late as 1988 the gaps were still wide. In Latin America 88 per cent of urban dwellers had a clean water supply, against only 57 per cent of rural residents. In Africa the urban to rural advantage was 74 per cent to 31 per cent. Only in South East Asia was provision almost equal, at around 70 per cent.[41]

Mortality is an indicator of overall quality of life. In eleven countries cited by the World Bank in its *World Development Report 1990*, infant mortality in rural areas was on average 57 per cent higher than in urban. In Bangladesh twenty- to twenty-four-year-olds in towns can expect to live 3·2 years longer than in rural areas.[42]

The livelihood gap

But the difference in livelihoods and incomes is decisive. In the 1960s urban incomes ranged from twice that of rural incomes in Latin America, to two and a half times higher in Asia and the Near East. In Africa they were eight times higher. Other data from twelve countries in the late 1960s and early 1970s showed similar results. Urban incomes in Africa ranged from four times rural in Côte d'Ivoire to almost nine times in Swaziland. Outside Africa the urban advantage was far lower, between two and two and a half times.[43]

To some extent, these differentials are inevitable. Manu-facturing and services can expand faster than agriculture. Demand for agricultural products usually grows only a little faster than population: people can only eat so much. Between

1965 and 1980 industrial production grew 142 per cent faster than agriculture in developing countries.[44]

Yet some of the differentials are the product of biases in government policy. In many countries high proportions of government jobs are in capital cities and a few large towns. In Lesotho, for example, two-thirds of Ministry of Agriculture staff are based in the capital, Maseru. Among those with degrees, the proportion is 93 per cent.[45]

These same few places also attract a disproportionate share of investment in infrastructure, industry and services. As a result their share of employment and incomes is often out of all proportion to their share in the population. Though it had only 15 per cent of the national population, Haiti's capital, Port au Prince, garnered 83 per cent of government current public expenditure. In 1970 São Paulo, with less than 9 per cent of Brazil's population, produced 48 per cent of the country's industrial production. Metro-Bangkok, with 11 per cent of Thailand's population, accounted for 37 per cent of gross national product.[46]

This is, of course, big-city bias rather than simply urban bias, as writers like David Satterthwaite and Jorge Hardoy have rightly pointed out. Many smaller cities and towns, dependent on the prosperity of their surrounding farm areas, also lose out when there is a bias against agriculture.[47]

And it is usually a bias to the rich and powerful, rather than to urban groups as a whole. Since the élite live in towns, élite bias shows up as urban bias. No one familiar with most Third World shanty towns could possibly imagine that governments entertain the slightest bias in their favour.

But the élite-serving bias does not stop at the élite. The benefits of unequal growth may not trickle down nationally. But within big urban areas, they usually do. Factories need workers, roads and modern flats need labourers, the rich need servants. And in their turn, these low-paid employees of the modern sector buy products and services from even lower-paid hairdressers, roast maize vendors, sellers of recycled cardboard, tea houses, beer halls, prostitutes.

Some areas of policy affect the terms of trade between agriculture and other sectors. The price of food, for example, is an expense item for urban dwellers, rich or poor. For farmers above subsistence level, it is income. Cheap food means higher real incomes for urban dwellers, lower incomes for farmers. Government marketing monopolies, crop price fixing, over-valued exchange rates, all keep farm incomes down, usually to the short-term advantage of urban areas.

The cities' pull, then, has very little to do with population growth. It has much more to do with changes in employment, and government policy on prices, taxes and the distribution of spending.

But the push from rural areas is just as crucial, and population growth has a powerful effect on rural incomes and livelihoods.

We saw in Chapters 9 and 10 how it can marginalize peasants. In Asia and Latin America, and increasingly in Africa, it reduces the size of farms and increases the numbers of landless and near landless. The rural labour supply increases, depressing rural wages. In some places this may be offset by higher demand for labour. Green revolution technology often demands more labour, to apply fertilizer and pesticide, transplant and weed more carefully. But in other places mechanization destroys rural livelihoods and aggravates the labour surplus. In Paraná state, Brazil, a shift from coffee into mechanized soybean production contributed to a net out-migration of 2·5 million people during the 1970s.[48]

And we saw in Chapters 8 to 11 how rural population growth can accelerate land degradation and force peasants to migrate to cities.

Is city-population growth the root of city environmental problems?

In cities, as in rural areas, population growth never acts alone. It always acts in conjunction with other factors which aggravate or alleviate its effects.

The sheer pace and size of population growth in cities poses some problems in itself. Among the world's 100 largest cities, Medan in Indonesia, Dhaka (Bangladesh) and the Nigerian capital, Lagos, grew at an average of more than 7 per cent a year for the fifteen years up to 1985 – doubling in size every decade. Bangalore, Ankara, Algiers, Pusan (in South Korea) and Tehran all grew at over 5 per cent per year, doubling in less than fourteen years. São Paulo grew by almost 500,000 people a year between 1970 and 1985, Seoul by 320,000. Few city administrations on earth, however committed, however rich, however efficient, could build roads, pipes, sewers and cables to cope with growth at this pace.[49]

But as with agriculture, technology is crucial, too. Decent and clean conditions could be provided, even for fast-growing urban populations, if low-cost technologies were used. Sites can be laid on for self-builders, providing road layout and a basic minimum of services. Community groups can be supported to upgrade their own shanty towns. Restrictions on cars and encouragement of buses can alleviate congestion, pollution and fare costs. Encouragement of small and informal businesses can create many more jobs than help for large-scale industry.

Yet all too often the limited funds that are available are spent on building housing to standards that the majority cannot afford. On expensive highway projects whose prime purpose is to provide road space for private cars. On large-scale capital-intensive industry.

Behind the choice of technology and the availability and distribution of spending lie other, political factors. In many developing countries power lies in the hands of an élite of big business, import and export traders, politicians, military and civil service élites. These groups ensure that their own needs are met first. They are often slumlords, unwilling to control rack rents or provide secure tenure. In many cities, especially in Latin America and the Near East, ample wealth exists to alleviate urban problems, but the political will to tax it has been lacking.[50]

Phases in city–environment relations

But population growth always plays a part in the dynamics of urban areas. For any given level of political bias for or against the poor, higher rates of population growth will mean worse environmental conditions within cities.

Here, too, there is a kind of adjustment process at work which controls environmental impact – but it is much more complex and unreliable than with natural resources, because it is mediated by the political system.

It is often surprisingly slow. London suffered the grime and gloom of smog for three centuries before coal-burning was banned. For even longer it tolerated the stench of the Thames, polluted with the filth of untreated sewage and later the poisonous effluents of factories. It drew its drinking water from this open drain until the Metropolitan Water Board was set up in 1903. These delays were due to the absence of affordable technologies, or to a lack of understanding of the link between filthy water and disease.[51]

Today the linkages and technologies are universally known, and when pollution impinges on the rich and powerful, it is promptly dealt with.

But when the poor are victims, action waits until they achieve political muscle through protest, riot or the ballot box. Higher rates of population growth do not speed up the process of pressure and response. Modern armed dictatorships are tenacious. Efficient repression can keep the lid on protest for a very long time, even under appalling conditions. There is an adjustment mechanism: over the long run, people will rebel and will eventually rebel successfully. But the process is jerky, unpredictable, and unreliable. And in the meantime a great deal of human suffering can be endured.

Democracy is an indispensable precondition, of course, but it is no guarantee of speedy corrective action, for in poor countries the illiterate rarely assert their full legal or political rights. Even in Western democracies the poor usually get the poisons. I remember, in the 1960s, Cardiff's steelworks used to shroud the sun and spew red dust all over adjacent Splott. Clean laundry was dirty before it had been worn. Splott, needless to say, was a poor working-class district.

So the process of adjustment is long indeed. And it works at first only inside cities, where population density is higher and the impact of pollution on people more intense. The city's own internal problems are dealt with first, starting with the biggest health risks.

And they are first dealt with, not by getting to the root, but by exporting the problem outside the city, to places where population density is lower. Street corners are cleared, but waste is tipped on city fringes, or in rural areas. City water supplies are cleaned up, but polluted water flows downstream. City air is clarified, but smoke is spewed across the countryside, or across national boundaries.

13

THE DROSSY AGE:
monuments in solid waste

Fast Runner destroyed four canoes of his father-in-law. The black bear blanket of Throw Away was scorched, and the skin of his legs was blistered, but still he held his ground.

Account of a Kwakiutl potlatch[1]

Towns and cities have their most devastating impact not through what they consume, but what they waste.

Rural settlements in non-industrial countries don't generate much waste. Food, building materials, fuel are collected locally. They don't have to come packaged. Wastes are mainly organic. They degrade easily, or can be easily recycled, as fodder for livestock, fuel or fertilizer.

But in cities, waste became a severe problem long before the time of plastic and aluminium. In modern Rome, the 49-metre Monte Testaccio is entirely composed of broken potsherds dumped by storekeepers from the emporia that lined the river Tiber in ancient times. Medieval Londoners dumped their sodden floor-straw out of the door, sometimes completely blocking the lanes. Butchers and fishmongers tossed offal and reeking fish into heaps.[2]

As the city grows its supply-lines lengthen. Energy is needed to bring goods in. More energy is used to take the remains away. Since the industrial revolution, that energy has been supplied mainly by burning fossil fuels and emitting air pollutants.

Almost everything the city consumes is supplied in a sheath of waste. Food supplies come packaged against damage, preserved against decay. Fuel arrives in barrels, bricks on pallets. In industrial countries packaging accounts for between a third and a half of household waste. The average American discards four times his or her own weight in packaging every year.[3]

Once urban technology and distribution systems take hold, rural areas become no more self-provisioning than cities. In terms of waste output and resource use, most rural residents in developed countries live urban life-styles.

Wealth boosts waste

As usual, population, consumption and technology are all crucial in determining the amount of waste produced. Population fixes the number of persons. The level of consumption decides the amount of goods each person uses. Technology governs how much waste is produced for a given quantity of goods.

Consumption of goods per person rises as incomes rise, and so does waste production. The old adage 'waste not, want not' is stood on its head. Those in want waste least. In poor countries, typical town dwellers produce only 0·5 kilos of municipal solid waste per person per day. In middle-income countries they generate about 0·7 kilos. In the late 1980s the daily production in high-income countries ranged from a low of 0·92 kilos per person in Western Europe right up to 2·4 kilos in America, the most wasteful nation on earth.[4]

Those who want not waste most. As incomes rise, people can afford to renew products, from cars to computers to clothes, more frequently. Manufacturers compete to improve product performance. Design and fashion change continually. Builders in wealthier districts of London tell tales of almost-new fitted kitchens, torn out and thrown out because the new owner disliked the colour. Old products are cast away rather than lovingly restored. As the value of time rises, people are less willing to spend it in repairing or salvaging.

The world pattern of manufacturing is shifting, and this, too, plays its part. More and more consumer durables are produced in the Third World where labour costs are low. High labour costs in Western countries makes repairs costly, and replacement more attractive. Fitting my two-year-old computer with a new motherboard cost more than buying a new machine of the same type.

The amount of packaging also rises as countries 'develop'. Packaging not only protects, but projects a brand image or provides consumer information. More packaged pre-cooked and take-away meals are eaten to save cooking time. In New York a single health-conscious salad bar lunch with coffee, juice and yoghurt produces enough waste to fill a paper bin: one paper bag, one aluminium can, one polystyrene cup and lid, plastic cutlery, paper tissues, plastic salad container, plastic yoghurt pot and foil lid.

The rich waste far more than the poor. Even within the same country waste output rises with income. Among the urban elites of Third World countries it rivals the highest Western levels. In Port Harcourt, Nigeria, for example, the output of domestic refuse ranges from only 0·55 kilos per day in Maintown and Borokiri districts to the above-US level of 3 kilos a day in Rumuema and Rumuokoro. In Cairo the resource value of household waste in upper-income households is almost three times that for low-income.[5]

The make-up of waste changes with economic growth. In poor countries most of it is bio-degradable. Between 40 and 85 per cent by weight is made up of vegetable matter, and the moisture content is high. Other organic materials such as wood, bone, straw, and leather are also significant.[6]

As incomes rise, the quantity of vegetable waste stays roughly constant, at between 0·3 and 0·4 kilos per day. But consumption of non-food goods grows, and technology increases the share of plastics and metals. So the quantity and proportion of waste that is not bio-degradable grows. In Lucknow, India, vegetable matter makes up four fifths of waste; in Brooklyn, New York only one fifth. Paper makes up 35 per cent in Brooklyn; only 2 per cent in Lucknow. Plastic and metal make up a quarter of Brooklyn's waste, seven times the proportion in Lucknow.[7]

The figures so far are for municipal waste alone. But industrial waste is far more massive. In the late 1980s it totalled 1430 million tonnes in OECD countries, against only 420 million tonnes for municipal waste. Output of industrial waste per person in Western countries ranged, in the late 1980s, from a low of 0·48 tonnes per person per year in Sweden right up to

2·54 tonnes in Japan and 3·05 tonnes in the United States. For Eastern Europe the figure was estimated at 1·3 tonnes. For developing countries, a mere 46 kilos – one-thirty-eighth as much as the average Westerner.[8]

Hazardous waste also grows with economic and industrial development. Western countries pump out 367 kilos per person every year. They put out no less than 89 per cent of the world total in the late 1980s. Eastern Europe and Russia, with 49 kilos per person, produced 6 per cent. The developing countries as a whole produced only 5 per cent of the world total – a mere 4 kilos per person. Each Westerner is responsible for over ninety times more hazardous waste than the typical Southerner.[9]

This does not exhaust the list of wastes. The OECD produces as much as 7000 million tonnes of mining spoil, demolition debris, dredge spoil, sewage sludge, residues from energy production and agriculture.[10]

Monuments in solid waste

The population contribution to the growth of waste varies from one country to another. Between the mid 1970s and late 1980s, for example, population growth accounted for only 16 per cent of the increase in municipal waste output in Japan. The rest was due to increases in consumption or changes in technology. In Western Europe the population share in growth was 31 per cent, and 38 per cent in North America.[11]

At any particular moment in time the role of population in waste output is simple. The total waste is the amount produced per person, multiplied by the number of persons. The more people, the more waste.

The calculus is one of the most depressing facets of environmental degradation. The most lasting monument all but the Shakespeares among us will leave is a huge spoil heap of refuse hundreds of times our own weight.

Even the poor have a big impact. Over the average lifetime of sixty-three years, a typical Third World city dweller will confer to the earth 149 times their bodyweight in combined municipal and industrial waste. But this is modest compared to a typical

European. After their allotted sixty-six years, they will bequeath 971 times their own bodyweight in debris to the biosphere. Piled into a cube this would measure eight metres tall and square – equal to the volume of the average small starter house.

Towering over the rest will be the memorial left by present-day North Americans. Each one will endow the globe with a vast mausoleum of litter 3900 times their own weight. This would form a cube more than fourteen metres on each side. No statistic better expresses the terrifying ecological impact of human activity on earth.[12]

If this deluge of filth is not dammed, the future implications are even more alarming. Let us peer forward to a world a century hence, when we may number eleven and a half billion. Assume that the average person by then produces waste at current European levels of one tonne per person. Spread out in a layer one metre deep, the annual refuse output would cover 77,000 square kilometres – an area roughly the size of Ireland or Sri Lanka. At that rate it would take only 1700 years to smother the entire land surface of the globe in garbage.

The disposal problem

As cities grow and non-organic wastes increase, open dumping with sporadic collection is the initial disposal method. Regular municipal collection is a luxury reserved at first for élite districts. The poor are left to tip where they can. Almost a quarter of Bangkok's waste is dumped on vacant areas or into canals and rivers. In Manila over 2000 cubic metres of rubbish are thrown into waterways every day.[13]

In this early phase, recycling is a significant employer and provider of cheap materials. The great social observer Henry Mayhew reported on a massive industry of scavenging in Victorian London which let almost nothing go to waste. Mudlarks collected coal and wood chips from the river at low tide. 'Shore-men' waded through sewers looking for lost coins under the filth. There were collectors of cigar-ends; buyers of used kitchen grease; and 'pure finders' who gathered dog-dirt, prized in the tanning of leather. In the early 1840s contractors used to pay local authorities for the right to collect ashes and refuse – the ashes could be sold to make bricks and condition marshy soils.[14]

Most poor families in developing countries today recycle before waste leaves the household. Other, poorer, families live by recycling what does get thrown out. In Africa, market stalls sell second-hand cardboard, bottles and tins. In the Andes pots and kettles are smithied out of old tin cans.

In Mexico City, some 10,000 scavengers work on the city dumps, and recycle a quarter of the city's waste. Manila has an estimated 5000 scavengers serving a population of 8 million. Lima's 1000 scavengers recycle paper, steel, glass and plastic saving up to $26 million of foreign exchange. Cairo would choke with refuse but for the Zabbaleen, a clan of Coptic Christians who collect wastes from high- and middle-income households. They sort the carts in their courtyards, rear pigs on the edibles and sell the rest to dealers in all kinds of materials.[15]

Recycling becomes much less prevalent as incomes rise. The more there is to recycle, the less recycling is done. It is poverty that promotes scavenging. Few people in developed countries are ready to face the stigma of sifting through rubbish dumps. Livestock no longer roam the streets eating vegetable wastes. Animals become a source of refuse, through pet food and litter, rather than a means of clearing it.

As municipal waste collection becomes the norm, large-scale movement of wastes begins. Landfill is the first recourse. Greece, Ireland, Hungary, Australia, Canada, Finland, the USA and Britain all dump more than 80 per cent of their wastes in landfill sites.

Old sites filled without due precautions are now posing massive pollution problems. As early as the first half of the 1970s West Germany had 50,000 contaminated sites, 5000 of them needing treatment. In 1988 one in four of the Netherlands' 4000 polluted sites needed immediate clean-up, at a cost of some $6 billion. In the USA 21,512 potentially hazardous sites had been identified in 1985. The cost of cleaning them up was estimated at between $20 and $100 billion.[16]

Landfill can cost between $20 and $60 per tonne in developed countries. But the cost and difficulty of landfill increases as space runs out and loads have to be transported further and further afield. Hazardous waste is often exported, dumped on someone else's doorstep. More than 2 million tonnes of hazardous waste move across European frontiers each year in 100,000 lorry trips. The former West Germany exported one million tonnes – 18 per cent of her total – in 1988, most of it to the German Democratic Republic.[17]

As landfill problems grow, incineration becomes more attractive. Denmark, France, Switzerland, Sweden all burn more than 30 per cent of their municipal wastes. Japan incinerated 73 per cent in 1987. But incineration, too, has its problems. It costs three to seven times more than landfill. Though high temperature treatment can reduce the volume to one-sixth of the original trash, it still leaves ash requiring disposal. It produces local air pollution, contributes to acid rain and global warming. It leads to local protests which demand pollution controls and bump up the costs.[18]

At this stage recycling comes back into fashion. It not only reduces the waste disposal problem. It also saves large amounts of energy and water, and reduces atmospheric and water pollution. Recycling aluminium uses 96 per cent less energy than refining it from bauxite. Recycled steel requires 74 per cent less energy to produce than steel from iron ore. Reuse of steel scrap also reduces water use by 40 per cent and water pollution by 76 per cent.[19]

Despite these advantages recycling is still only in its infancy. Recycled aluminium rose from 21 per cent of world consumption in 1971 to 30 per cent in 1987. The price of energy was the key factor here. Recycling spurted after the two oil-price hikes of 1973–4 and 1979–81, but each spurt was followed by a plateau.[20]

Recycled waste paper rose slowly from an estimated 20 per cent of world consumption in 1965 to 25 per cent in 1980. In OECD countries the average recovery rate for paper and board rose from 27 per cent to 34 per cent between 1975 and the late 1980s. For glass it rose from 22 per cent in 1980 to 32 per cent in 1990.

Progress in recycling has been very uneven between countries, showing the key influence of government policy. Some, like the UK and Sweden, were recycling less aluminium in 1987 than in 1971. Japan recycles 40 per cent of her aluminium, against only 19 per cent in the UK. Between 1976 and 1987 the proportion of steel recycled fell quite steeply in the USA and UK.

Reuse and repair create less waste and use fewer resources even than recycling. A refillable glass bottle, reused ten times, uses 24 per cent less energy than a bottle, made of recycled glass, but used only once. Yet reuse has not really begun on any scale. Indeed it has retreated. Returnable glass bottles once dominated the US drinks market. But by 1985 they had dropped to only 16 per cent. Aluminium cans and plastic bottles had captured 69 per cent of the market. In 1985 no less than 66 billion aluminium beverage cans were used in America – 275 for every person.[21]

The soundest option for cutting waste is reduction at source: use less packaging, reduce unnecessary consumption, make do with enough. But as the OECD's *State of the Environment Report 1991* points out, 'The move from theory to practice has yet to be made when it comes to the prevention of waste formation.'

Instead, municipal waste output per person continues to grow everywhere in the world. In Western countries municipal waste output per person increased by 1·3 per cent a year in the thirteen years from 1975. Even in Japan, which has one of the best records on recycling, it rose by 0·8 per cent a year.[22]

The evolution of waste management

As in other spheres, population growth plays a major role in driving the changes in waste management. The key stimulus is what might be called 'waste density' – population density multiplied by waste output per person.

In the first phase, populations are widely dispersed. Consumption per person is low. Waste is mainly organic. What cannot be naturally recycled is scattered in the open. There is so little that this does not create major problems.

In the second phase populations concentrate in towns and cities. The size of the waste problem increases with the size of the city. In pre-modern cities human and animal scavengers recycle much of the waste. In modern times, consumption of goods per person rises, and technology increases the proportion of inorganics. This is the phase of open dumping. It persists as long as people tolerate it – or for as long as undemocratic political systems prevent them from protesting about it. Eventually waste density makes local dumping impossible.

Here we enter the third phase, which is one of crisis and progressive adjustment. As waste density rises, disposal gets more and more expensive and problematic. At first, urban waste is transported to landfill sites in rural areas, where population density is lower. Gradually potential sites grow scarce, and incineration is increasingly used.

Most countries are still only on the fringe of the fourth phase: sustainable waste management. This is the phase of the four 'R's: repair, reuse, recycling and waste reduction. It represents, paradoxically, a return to the habits of more frugal days – though without the human degradation involved.

Countries with the highest population density are most likely to reach this last phase first. It is significant that Japan, with 330 people per square kilometre, is a world leader in progress towards incineration and recycling. The USA, with only 2·7 people per square kilometre, lags far behind.

This is not a recommendation of high population density. At every stage it is the problems that prompt the search for solutions. But problems are by nature uncomfortable. Solutions are rarely adopted before a great deal of damage has been done. Much of it will be expensive to remedy, much of it irreversible.

14

A SEA OF TROUBLES:
polluted waters

Waste is not just what we consciously throw away. It includes every by-product of our activities of which we make no further use. It embraces liquids and gases, as well as solids. Indeed the more volatile the material, the more widespread are its effects, the more difficult to control.

Liquid and gaseous wastes will test to breaking point human power to adapt resource management to population pressure. Their mobility is the heart of the problem. They spread far beyond the place where they are emitted. Through river systems and the circulation of the atmosphere, they can reach the farthest corners of the earth.

The great global commons in which both end their lives, the oceans and atmosphere, suffer all the problems of land-based commons. Everyone has right of access. No one enjoys ownership and control. Each individual, company or nation enjoys the full benefit of their own cheap disposal of waste – but suffers only a fraction of the pollution that results.

The degradation of a land-based common impinges on the lives of those who live near it and depend on it. Eventually they may agree to control its use – or put pressure on their local or national government to do so.

But the global commons are much harder to govern and police. Rivers, oceans and atmosphere respect no political boundaries. Effects may be felt hundreds or thousands of kilometres from causes, in another town, another country, another continent. The polluted may have no direct political way to exert pressure on pollutors.

Pure, clear water is a thing of the past. There are no pristine bodies of water left anywhere in the world, even in the remotest areas – that is the view of the United Nations Environment Programme's authoritative review of global freshwater quality. Even in the Arctic traces of toxic chemicals – polychlorinated biphenyls, DDT, lindane, lead, cadmium – have been found in air, seawater, sediments, fish, mammals and seabirds.[1]

Water-pollution problems are intimately linked to population density and urbanization. In rural areas disposal of excreta poses problems for human health, but not for the natural environment. It is low in volume, dispersed, and quickly recycled. There is little net loss of nutrients. The nitrogen cycle is closed and sustainable.

Urbanization removes people from proximity to the fields. In small rural towns, or on the fringes of cities, night-soil still can be composted and used as manure. Night-soil still provides one-third of China's fertilizer needs.[2]

But in most countries human excreta from urban areas are lost to the farming system. As farming intensifies, even livestock are reared in 'urban' conditions, and feedlot effluents pollute surface and ground water. In the Netherlands, 52 per cent of pig manure is wasted, 60 per cent of fattening calves' and 80 per cent of chickens'.[3]

Human and animal excreta turn from a resource, into a threat to life in rivers and coast. The land's nutrients are exported into the sea, and the nitrogen cycle is decisively broken. The loss is made up by using chemical fertilizers – which become another source of pollution.

Sewers concentrate sewage and dump it into waterways. The proportion of the world's population with sewer connections is steadily rising. In OECD countries it rose from only 33 per cent in 1970 to 60 per cent in the late 1980s. By the late 1980s, 13 per cent of Africans had sewer connections, 38 per cent in the Eastern Mediterranean, and 51 per cent in the Americas.[4]

There are still large numbers of urban people without sewers or latrines. But their liquid wastes may still go to pollute waterways. River banks, gutters, open ditches, or beaches are favourite places for excretion. In Jakarta enterprising shanty dwellers build latrines with plank walkways out over canals. Children play and bathe in what are open sewers.

Most urban sewage in developing countries is untreated. In Latin America less than 2 per cent of urban flows receive treatment. Treatment facilities in India cover less than a third of the urban population. Only 15 per cent of waste water in China receives treatment.[5]

Where it exists, treatment is limited in most cases to removal of solids. Even in developed countries it rarely goes beyond secondary treatment. This removes substances likely to deprive fish of oxygen. Tertiary treatment removes most pollutants. But in the late 1980s this covered more than a third of the population only in Finland, the former West Germany, Sweden and Switzerland.[6]

Some sewage solids are recycled. Most developed countries use about half their sludge on farmland, as fertilizer and soil conditioner. The other half becomes not an asset but a liability, dumped at sea, in landfills, or incinerated. As the size of cities increases, so does the cost of transporting sludge to farmers' fields over increasing distances.[7]

Waterways also receive industrial effluents. These have been increasing rapidly worldwide. In the mid 1980s around 237 billion tonnes of industrial waste waters were pumped out. By the year 2000 the figure is expected to almost double to 468 billion tonnes.[8]

Much water pollution originates from rainwater running off the land. By weight, the biggest water pollutant is sediment. Eroded topsoil, lost to the land, becomes an unwanted burden for rivers and seas. An estimated 13·5 billion tonnes of sediment reach the oceans every year. Only a third of this is natural. Human intervention is responsible for the other two-thirds.[9]

Fertilizers and pesticides are next in importance. In the Netherlands, about one-tenth of the nitrogen and one-quarter of the phosphates applied to the land accumulates in the environment. Global wastage is probably higher, especially in the tropics. Torrential rains can wash away high proportions of fertilizer.[10]

World fertilizer use has grown fast, averaging 6 per cent a year since 1961. Growth in developing countries was faster still, at 11 per cent a year. The total quantities of nutrients applied to the soil worldwide rose almost fivefold between 1961 and 1988, from 31 million tonnes to 146 million.[11]

Overuse of fertilizer contributes to one of the most widespread freshwater problems: eutrophication. This occurs when plants and algae are stimulated into excess growth by an overload of nitrogen and phosphorous from fertilizers, sewage and detergents. When the plants decay, oxygen in the water is depleted, and water animals die.

Two-thirds of lakes and reservoirs in eighteen western countries were eutrophic in 1982. The problem affected twenty-eight out of sixty-five lakes in Italy, and thirty-nine out of seventy-five in the former West Germany. A quarter of China's lakes are classified as eutrophic. In eleven Latin American countries 56 per cent of lakes and reservoirs are affected. The global incidence is thought to be around 30–40 per cent.[12]

Coasts

Half the world's population lives on the coast, another quarter within 60 kilometres of the sea. Hence pollution is particularly concentrated along coasts – and coastal pollution has an impact on large numbers of people.

Coastal areas are crucial to marine production. Two-thirds of all fish caught at sea hatch out in tidal areas. Measured in terms of carbon fixation, coastal mangrove forests, seagrass beds and coral reefs are as productive as tropical rainforests – and twenty times more productive than the open ocean.[13]

Eutrophication affects coastal as well as inland waters. It is reported from the Baltic to Tokyo Bay, from Long Island Sound to the mouth of the Amazon. It often takes the form of 'red tides' – blooms of phytoplankton which damage fisheries, hit tourism, and poison seafood. These have been increasing in frequency in many parts of the world. In Japan's Seto Inland Sea over 100 red tides occur every year. In Hong Kong's Tolo Harbour red tides increased from average of 2·3 a year in 1977–82 to 18 a year in 1983–7. In 1988 tides of algal sludge suffocated the canals and lagoon of Venice, and plagued tourists along a 1000 kilometre stretch of the upper Adriatic coast.[14]

Coastal wetlands, covering some 6 per cent of the world's surface, divide the dry land from the sea. They play an important role not only in fish-spawning, but as a buffer for the land against seawater floods and cyclones, and a buffer for the sea against sediment and pollution from the land. Since 1900 the world may have lost half its wetlands to drainage for agriculture, clearance for forestry, urban and tourist development. Asia is thought to have lost as much as 60 per cent of its original wetland area, Africa almost 30 per cent.[15]

The world's 240,000 square kilometres of coastal mangrove forests have been under severe attack. Malaysia has lost half its mangrove area in the past twenty years. Much was cut for conversion into wood chips for the Japanese market. More than a quarter of Indonesia's 2·5 million hectares of mangrove forests were felled and drained to grow rice, or rear shrimps. By 1982 half the Philippines' mangrove area had been converted into ponds for prawns and milkfish.

In all, Asia may have lost 58 per cent of its mangroves, Africa 55 per cent. Mangroves in the deltas of the Niger, Indus and Ganges have been severely damaged by upstream irrigation, which has drawn off much of the river flows and allowed salt seawater to intrude further inland.[16]

Coral reefs extend over a total area of 600,000 square kilometres, and protect 15 per cent of the world's coastlines. Like humans, corals modify the environment on a large scale for their own benefit. From space the Great Barrier Reef is still the largest visible structure created by any living organism.

In the process coral make favourable environments for many other species. One-third of all ocean fish species live here. Fishermen depending on coral-reef communities supply up to 90 per cent of fish production in Indonesia and 55 per cent in the Philippines.[17]

Coral reefs harbour the greatest biological diversity of any habitat after tropical rainforests. The reasons are similar: warmth, constancy of food supply, multiplicity of layers. But they face even more serious threats than the forests. Two-thirds of the reefs in a 1982 Philippines survey had less than 50 per cent cover of live coral. Less than 6 per cent were in good condition. Reefs fringing the south coast of Hainan Island in the South China Sea have lost 95 per cent of their live coral.[18]

We have seen the effects of dynamite fishing and coral collection (see p. 66). Pollution from agriculture, industry and sewage is a major threat. Eroded soil often ends up here, smothering the polyps.

Oceans

The open oceans, furthest removed from human activities, have so far been least affected. But the writing is on the wall.

Between 6 and 7 million tonnes of industrial waste were dumped at sea every year between 1975 and 1985. In the early eighties an annual average of 15 million tonnes of sewage sludge was dumped. The rate of oil spillages has declined significantly, but accidents and sluicing still spill some 1·5 million tonnes of oil into the oceans every year. Another 1·7 million tonnes come from the land via rivers and the atmosphere. To date some 60,000 Tera-Becquerels of radioactive waste have been dumped in the oceans, along with around 20 billion Tera-Becquerels from atmospheric nuclear tests. The total amounts to around a 1 per cent increase in the level of radioactivity naturally present in seawater.[19]

The most visible impact of human activities on the oceans is plastics. After only a few decades of widespread use, plastic is accumulating in terrifying quantities. A 1988 beach clean-up in the United States removed 907 tonnes of refuse from 5600 kilometres of shoreline – an average of 16 kilos per 100 metres. Some 60 per cent of it was plastic. One survey of sea-floor sediment, off the coast of the United Kingdom, found an average of 2000 pieces of plastic debris per square metre. In 1985 almost half a million plastic containers were dumped overboard from the world's shipping fleet.[20]

Plastic affects animal life too. In a recent study, plastic particles were found in the digestive tracts of 25 per cent of the world's seabird species. Nine out of ten Laysan albatross chicks examined on Midway Island had plastic particles in their digestive tracts. Every year plastics kill an estimated 1 million seabirds and 100,000 seals and cetaceans. Seals get jammed in plastic packing bands which cut into their flesh as they grow.

Turtles and whales are killed by eating plastic bags and sheets. At least 150,000 tonnes of plastic fishing nets are abandoned or lost each year and continue to 'ghost-fish' for years after, entrapping and killing fish for no purpose.[21]

The population connection

Clearly we are dealing with a very wide range of threats. Industrial effluents are the product of technology and economic growth. The rise in plastics is a change in technology and patterns of consumption.

But much of the environmental impact of water pollution is due to growing populations: the direct effect of the search for protein and livelihoods, and the indirect side effects of agriculture and urbanization.

The output of liquid wastes depends on population size, levels of consumption per person, and pollution per unit of consumption. All three factors are inseparably part of the equation. The relative contributions vary from one field to another.

In the area of domestic sewage, consumption is not a significant factor. Excreta increases with the amount of fibre consumed, but this makes little difference to environmental impact. The total quantity of sewage produced is a direct function of population size. The impact depends on population density and disposal methods.

Most of the increase in fertilizer use appears to be due to a change in technology rather than population increase. If fertilizer use per person had remained static, world fertilizer use would have increased by only 1·9 per cent a year over the twenty-seven years up to 1988. In fact, it increased by 5·9 per cent. In both developing and developed countries, population growth accounted for only just over one fifth of the increase, while increased consumption of agricultural products accounted for 8–18 per cent. The remainder was due to changes in technology (see appendix pp. 306–9).

But things are not so simple. Expansion in cropland has been slowing down (see pp. 44–6). Fertilizer use increased as a substitute for land to maintain food production. It can be argued that the increase was almost entirely due to population growth.

A tale of two rivers

River quality in most developed countries has followed a U-shaped curve. Freshwater quality touched bottom around the 1950s and has been recovering since then. Controls have been introduced on domestic and industrial effluents. These are being continually tightened. Farm chemicals are just beginning to be controlled.

The Thames's story is typical. In the eighteenth century the river had great quantities of salmon. After that water closets and sewers dumped their untreated waste. In summer the river reeked of rotting eggs – so powerfully, in the Great Stink of 1858, that the windows of Parliament had to be draped with curtains soaked in chloride of lime to allow members to breathe. Drinking water was contaminated. Cholera outbreaks grew in frequency. To prevent these, sewage treatment started around 1890 – but industrial effluents continued. By 1949 the river was devoid of oxygen over a twenty- to thirty-mile stretch. A survey in 1957 found no fish at all between Richmond and Gravesend.[22]

By 1974 two purification plants had been fitted with the latest treatment equipment, and the oxygen content of the water began to rise. Large numbers of whiting and smelt appeared. Invertebrate diversity increased and waterfowl returned. By 1979 almost 100 species of fish had been recorded. In 1983 the first salmon in a century and a half was caught.[23]

In most developing countries, water quality has been deteriorating rapidly. Out of sixty developing countries that are industrializing, only ten have effective laws, regulations and enforcement procedures to cope with growing pollution problems.[24]

Consider the Ganges. In the early 1980s India's holiest river was still plunging towards the bottom of the pollution abyss. Some 114 cities pour in their libations of untreated sewage. The Yamuna tributary, as it passes through Delhi, picks up a daily 200 million litres of sewage and 20 million litres of industrial effluents. These include half a million litres of toxic DDT wastes. Downstream, at the industrial city of Kanpur, only three factories out of 647 have treatment plants. The rest disgorge 200 million litres of effluents a day straight into the river.[25]

By the time it reaches Varanasi (Benares), the river has acquired a burden of silt from the rice fields, and is the colour of strong tea with milk. Sixty-seven drains spew almost 60 million litres of untreated sewage into the river every day. Every year some 10,000 unburned corpses and 60,000 animal carcasses are dumped into the river. Here 6 million pilgrims a year immerse themselves in the river's spiritually cleansing, physically filthy waters. Water surveys have found numberless cholera, dysentery and typhoid germs. An estimated 98 per cent of the city's residents suffer from stomach ailments. I remember Varanasi for the splendour of the bathing ghats at dawn, for boatmen tipping the shrouded corpse of a smallpox victim overboard – and for a day and a night spent rushing between sickbed and toilet.

Lower still, the Hooghly estuary – one of the many mouths through which the Ganges reaches the sea – is choked by untreated industrial wastes from more than 150 factories around Calcutta, where raw sewage pours into the river from 361 outfalls.

Management phases

The management of liquid wastes, like that of solids, follows the same stages driven by increasing population and pollution density.

Scattering by dispersed rural populations creates few problems, except to human health. There may be limited social controls on where people may excrete.

When people gather in urban settlements, large-scale dumping of liquid wastes into waterways begins. As industrialization follows, industrial effluents are added to domestic.

Crisis comes as the increasing density of pollution creates a rising level of health, fishery, amenity and other problems. At the nadir waterways reach a state of degradation which stimulates widespread public concern. Where the political system permits, protest and official action follows.

The transition to sustainable management follows. The most serious threat to human health, sewage, is dealt with first – with a gradual increase in the number of pollutants removed before discharging. Industrial effluents are controlled later. As leisure increases, controls extend to pollutants that threaten wildlife. The state of rivers improves.

Effective controls on water pollution usually begin only when problems have reached a fairly severe level. Democratic processes must also be strong enough to allow the public to protest, and to induce politicians to respond. These two factors explain why water pollution, like air pollution, is worst in newly industrializing countries.

As the impact of pollution spreads, there is a gradual progression from localized measures to control over wider and wider areas. Coastal pollution in many countries has already reached the stage where controls on its land-based causes have begun. Limits on use of fertilizers and pesticides will soon be common, in developed countries.

Ocean controls are modest so far, largely limited to dumping by ships of their own or other people's oil, sewage and solid wastes. Yet ships are responsible for less than a quarter of ocean pollution.[26]

The oceans, like the other fluid global common the atmosphere, will be last to be effectively protected.

15

A CONGREGATION OF VAPORS:
air pollution and climate change

> The roof of the house caught fire and the whole
> house was nearly destroyed, while those who were
> concerned kept their places with assumed indiffer-
> ence and sent for more possessions to heap upon the
> fire.
> Account of a Kwakiutl potlatch where 400 blankets
> were burned[1]

Throughout history the human race has expanded its numbers
and its demands on the environment, over an ever-increasing
area of the globe. At certain stages, we have come up against
limits. Limits of wild cereals, limits of wood, limits of waste
output.

Now, in the last decade of the twentieth century we are
expanding faster than ever before, into the remotest ecological
zones. We are pushing against the outer limits.

The atmosphere is our final frontier. Our capacity to change
the air above is perhaps the most telling symbol of human
hubris.

The burning rains

We have the awesome power to wound Gaia severely. It is only
recently that the earliest scars became visible.

From the 1960s signs of ill health first appeared in the trees of
Germany's Black Forest: discoloration, then shedding of leaves
and needles; thinning of crowns and canopies; and a decreased
resistance to frost, drought and disease. By the late 1970s a

third of the forest's firs were dead, and symptoms had begun to appear on spruce, oak and birch.

Forest death is the outcome of a complex of causes which are still not completely understood. Acid deposition – through rain, snow, mist and dew – is a key player. It mobilizes poisonous metals like aluminium in the soil, and induces a deficiency of calcium and magnesium.

The major causes of acid rain are gases emitted by burning fossil fuels: sulphur dioxide and oxides of nitrogen. These combine with moisture not just in rain or snow, but also in fog or dew on the ground, to form sulphuric and nitric acids.

Germany was only the first to show symptoms. The illness quickly became an epidemic. By 1989 more than half the trees surveyed in Czechoslovakia, the United Kingdom, Germany, Estonia and Tuscany had lost over 10 per cent of their leaves. In Czechoslovakia and the United Kingdom, one in four trees had lost more than a quarter of their leaves. In the United States, three out of five trees had lost more than 10 per cent of their leaves. One in five had lost over a quarter.[2]

Acidity is measured on the pH scale. A value of 7 is neutral. Anything above that is alkaline, anything below is acid. Natural rainfall picks up some acidity from nitrogen in the air, and has a pH of around 5·6. In many parts of Europe and the USA rainfall has clocked pH values of 4·5 or less. Since the pH scale is logarithmic – every one-point change means a tenfold increase or decrease – these values are over ten times more acid than natural rainfall.[3]

Acid rain damages many forms of wildlife. Salmon cannot tolerate pH levels much below 6. Perch die out below 5. In Sweden fish stock has declined in 9000 lakes. In Norway seven out of every ten lakes are acidified: fish have disappeared from an area of 13,000 square kilometres. Birds that depend on coniferous forest, like the goldcrest, or on water, like dipper, osprey, red-throated diver or black duck, have declined in numbers.[4]

Many developing countries are beginning to suffer. In Jakarta, Indonesia, rainfall recorded pH values of 4·6. In a wide area around Chongquing and Guiyang in China, levels below 5 have been reached. Large areas of paddy rice have turned yellow, and as far as twenty kilometres from Chongquing stands of Masson pine have withered. In Brazil pollution from the industrial city of Cubatão has damaged montane rainforest in Serra do Mar.[5]

Human output of sulphur dioxide now rivals that from natural sources like volcanoes. Even the huge eruption of Mount St Helens emitted the equivalent of only 1 per cent of US emissions from human sources. Global emissions of sulphur dioxide from human activities rose from less than 10 million tonnes in 1860 to an estimated 160 million tonnes in 1980.[6]

Trends are now diverging between North and South. In the developed countries sulphur dioxide was one of the first air pollutants to be controlled. A shift to fuels with lower sulphur content, coupled with strict requirements on flue extraction, cut sulphur dioxide emissions from OECD countries by 38 per cent between 1970 and 1988. Some countries performed even better. Emissions in Sweden fell by 79 per cent, in Japan by 83 per cent. Some countries still have a long way to go. Canada put out a massive 146 kilos per head in 1988, the USA 84 kilos and the UK 63. Each Swiss person, by contrast, emitted only 9·4 kilos and each Japanese a mere 6·8.[7]

Few data are available for developing countries. But all the pointers suggest that their output of sulphur dioxide has been increasing in pace with the spread of industrialization. According to one estimate, while Northern emissions grew by 17 per cent during the 1970s, Southern output rose by 45 per cent. Asia's increase was no less than 53 per cent. In 1981–3 China's emissions were third after the United States' and the USSR's. Air pollution in cities provides another measure: in 1980–4, out of twenty-seven cities with sulphur dioxide levels higher than World Health Organization guidelines, sixteen were in developing countries.[8]

The other principal acid rain culprits, nitrogen oxides, are produced mainly by high temperature combustion, which combines nitrogen and oxygen in air. About half the nitrogen oxides emitted in Common Market countries comes from vehicle engines, and much of the rest from electricity generation. Attempts to limit car emissions have been modest: reductions for each individual car have been more than wiped out by the rise in car numbers. During the 1970s emissions of nitrogen oxides in the OECD rose by 17 per cent. In the seven

years after 1980 they fell slightly, by 4 per cent, due mainly to
the rise in oil prices. Here, too, emissions vary wildly between
countries, from only 9·6 kilos per person in Japan in 1987, right
up to 80·4 kilos in the USA.[9]

Acid rain is thought of as a disease of industrialization. Yet it
can also be produced by the burning of forests and grasslands.
In Africa pastoralists regularly torch the savannahs at the tail
end of the lean season, to keep down shrubs and promote a
flush of new grass for their herds. Every year some 440 million
hectares burn, carrying formic, acetic and nitric acids, plus
ozone and hydrocarbons, over long distances. Over the forests
of north Congo and the Côte d'Ivoire, highly acid rain has been
measured with pH values of 4·4 to 4·6.[10]

Acid rain is a time-bomb for the Third World. As
industrialization advances, it will compound the damage of
deforestation. Many tropical zones are especially vulnerable.
Vast areas have laterite soils, highly leached by millions of
years of heavy rainfall. These are already acid, low in calcium,
with high concentrations of aluminium and iron oxides. Plants
and crops that have evolved tolerance may not be able to
absorb further increases in acidity.[11]

The hole in the sky

Forest death was the first great shock. The ozone hole appeared
only a few years later.

The thin ozone layer in the stratosphere is like a barrier
cream for life, filtering out the most damaging forms of
ultraviolet radiation. In the early 1970s there were fears that the
layer could be damaged by supersonic flights, space shuttles or
nitrous oxide emitted by fertilizers.

The real danger came from the most unlikely direction.
Chlorofluorocarbons were invented in 1928, and first used in
fridges and air conditioners. Later they were harnessed as
cleaning solvents, to propel aerosols and to puff up polystyrene
foam for cups and hamburger cartons. Global production of the
two main chlorofluorocarbons, CFC-11 and CFC-12 soared,
from a mere 2200 tonnes in 1940, to 126,700 tonnes a decade

later and 491,700 tonnes in 1970. The average growth rate in output over those three decades was a dizzy 20 per cent per year. Most of this was in the North. Developing countries produced only 14 per cent of 1986 emissions.[12]

Inert, stable, non-flammable, non-poisonous, CFCs seemed totally harmless to the biosphere. But because of their chemical stability, they are very longlived. They can persist for up to 130 years. And so they steadily accumulated in the atmosphere, increasing by 4·5 per cent a year between 1970 and 1988.[13]

The first caution came in 1973, when Professor Sherwood Rowland at Berkeley University, studying their potential impact on the atmosphere, was horrified by his own results. The United States began phasing them out four years later, followed in 1980 by the EEC. World CFC output fell at first, but then began to rise again.

Then, in 1982, British scientist Dr Joe Farman, taking measurements from Halley Bay in the Antarctic, first discovered the ozone hole. Over the next two years the hole grew bigger. Farman's results were published in May 1985. By October 1987 ozone levels around the Antarctic were less than half their levels in 1970. Two years later the hole covered 14 million square kilometres. Global surveys confirmed that ozone was being lost in all regions – ozone levels over Europe in 1991 were 8 per cent down on a decade earlier.[14]

. The potential impacts were drastic. If CFC use had continued .to grow, by the year 2075 there could have been an additional 43 million cases of cataract, plus 98 million cases of skin cancer, including 2 million extra cancer deaths.

Enhanced ultraviolet radiation also cuts the yields of many plants – from 5 per cent for wheat up to 90 per cent for squash. It reduces plant height and leaf area, and reduces flowering and germination. It damages cyanobacteria which fix nitrogen in paddy fields, and could slash rice yields. It could speed global warming by damaging trees and phytoplankton, which absorb large amounts of carbon dioxide.[15]

The overall costs of uncontrolled ozone depletion to the USA alone would have reached $150 billion by AD 2075, including lost crop harvests of $42 billion and fish catches worth $7 billion. Indeed, if skin cancer deaths are accounted at typical compensation rates, the total costs rose to $3517 billion.[16]

One woe doth tread upon another's heel

A third shock came close behind the second.

Concern about future global warming had been soberly expressed in scientific quarters. Then, during the 1980s, climatic records of all kinds were toppled across the world. Extreme drought hit the Sahel. Severe gales wreaked havoc in Britain's woodlands. The 1988 drought in North America's corn belt convinced many that global warming was already upon us.

Average surface air temperatures have increased by 0·3–0·6°C over the past century. Melting glaciers have raised the average sea level by 10 or 20 centimetres. The five warmest years on record were in the 1980s. These changes fall within the range of natural variability. But more and more scientists accept that they are symptoms of the impact of human activities on the global climate.[17]

The greenhouse effect has existed as long as the earth has had an atmosphere. Briefly, the wavelength at which energy is radiated depends on the temperature of the source. The hot sun radiates at short wavelengths. These pass easily into the atmosphere, as they do through the glass in a greenhouse. But the cooler planet surface radiates some of this energy back at longer wavelengths. These are not transmitted so readily, and some of the energy is trapped by gases in the atmosphere, as it is by a greenhouse, keeping it warmer than otherwise.

Venus and Mars both have their own greenhouse effect. Mars's version is too cool for life, Venus's is too hot. The earth's, like baby bear's porridge, is just right. Without it the average temperature on earth would be minus 18°C, colder than a Moscow winter, instead of the 15°C of a London springtime.[18]

Global warming involves an unnatural stoking of the natural greenhouse effect. It is happening because, like the cyanobacteria before us, we are altering the chemical make-up of the atmosphere.

Carbon dioxide accounts for 55–60 per cent of the warming effect. Like water and nitrogen, carbon has its own vast natural cycles. Some 39,000 billion tonnes are stored in the oceans. Another 1,550 billion tonnes are closeted in soils. Land plants stock 560 billion tonnes – 445 billion tonnes of that locked up in forest and woodland. Soil and plants give out and reabsorb around 100 million tonnes per year.[19]

Next to these gigantic numbers, human output of carbon dioxide is still dwarfed. Fossil fuel burning emits only around 5·4 billion tonnes of carbon each year. Tropical deforestation contributes perhaps 1·6 billion tonnes. Oceans and vegetation between them absorb just over half of our output – leaving 3·4 billion tonnes to boost the 750 billion tonnes of carbon dioxide in the atmosphere.[20]

But our small annual addition accumulates over time. By 1990 we had increased the carbon dioxide concentration in the atmosphere by a quarter over pre-industrial times, from about 280 to 353 parts per million. The level is climbing at a rate of 0·5 per cent a year.[21]

Second in impact are the chlorofluorocarbons. They not only destroy the ozone layer: they are also potent greenhouse gases, with 3500–7300 times more warming power, weight for weight, than carbon dioxide. In the 1980s they were responsible for around a quarter of the man-made greenhouse effect.[22]

When CFCs have been phased out, methane will become the second most important greenhouse gas. Man-made methane accounts currently for around 15 per cent of global warming. Although it breaks down after only ten years or so in the atmosphere, over the duration of a century each kilogramme is twenty times more potent than a kilo of carbon dioxide. Atmospheric levels of methane have more than doubled since 1800, from 800 to 1720 parts per billion. The level is increasing by 0·9 per cent each year.[23]

As with carbon dioxide, methane emission is also part of the natural cycle. Some 170 million tonnes per year are produced by wetlands, termites, oceans and freshwater. But man-made output, running at 360 million tonnes a year, is now double that of natural sources. Rice paddies bubble out some 110 million tonnes per year. The flatulence of livestock comes a near second, venting an annual 80 million tonnes. Burning of forests and grasslands release 40 million tonnes. Gas-drilling and coal-mining give off 80 million tonnes, while decomposing rubbish tips produce another 40 million tonnes.[24]

The fourth major greenhouse contributor is nitrous oxide. Natural emissions total around 6 billion tonnes. Human output as yet is only a quarter of that. The chemical breakdown of fertilizers is thought to be responsible for four-fifths of this.

Turning up the heat

Projecting current trends, the International Panel on Climate Change suggests that global average temperatures could rise by an average of 0·3°C each decade over the next century. By 2025 they could be 1°C warmer than now. By the year 2100 they could have risen by 3°C. Half-way through the twenty-first century, temperatures will be higher than at any time in the past 150,000 years. The speed of increase will be unprecedented: fifteen to forty times faster than at the end of the last glaciation.[25, 26]

Many uncertainties still remain. Cloud formation may increase, as more water is evaporated from oceans. This may moderate global warming, but the effect will depend on the type of cloud. There may be shifts in large-scale ocean currents, which are not well understood as yet.

The overall effect of feedback mechanisms is unresolved. Some environmental changes may dampen global warming. Deforestation and desertification increase the albedo or reflectivity of the earth's surface, so that less solar radiation is absorbed. Sulphur dioxide emissions increase the reflectivity of clouds. Global warming itself may bring on moderating changes. There may be an overall increase in plant growth, which will take up more carbon dioxide.

But most feedback effects seem likely to speed rather than brake the warming. Depletion of the ozone layer will damage plants, trees and plankton and reduce their carbon uptake. Global warming itself seems likely to increase deforestation. It will accelerate the decomposition of carbon-storing organic matter in the soil – the Arctic tundra alone has 160 billion tonnes of carbon locked in its frozen soils, some of which could be released with a thaw. Methane output from peat bogs and tundra soils could increase. So would water vapour in the atmosphere – and water vapour is itself a greenhouse gas.

On balance these complex interactions and feedback loops seem likely to step up global warming, rather than slow it down. The rise in global temperatures could then approach the upper end of the IPCC estimates – over 5°C by 2100.[27]

The earth's climate does not always change smoothly. At certain points slow quantitative changes pass thresholds at which they accelerate suddenly and catastrophically, like a ball rolling along a field and reaching a cliff edge. Slight changes in the earth's tilt and orbit, or massive series of volcanic eruptions, trigger ice ages after only a few degrees change in average temperatures.

We are like novices riding a capricious horse: we don't yet know how to read its moods, how to predict when it might decide to throw us.

Living with change

We do not even know, as yet, whether the net effects of global warming will be negative or positive. Previous warm periods have brought mixed blessings. During the late Pliocene epoch, around 4 million years ago, middle-latitude temperatures were 4°C warmer than today. The Sahel, the Sahara and Central Asia deserts had more rainfall. South-eastern England was subtropical, and northern Iceland temperate. But the sea level was a good deal higher.[28]

Past warm spells arose from natural causes, not artificial ones, so they cannot serve as reliable guides to the future. When we talk about the impact of global warming, we are groping in the dark. We have no certain view of how everything hangs together, only scattered and incomplete impressions. What we do know is that there will be multiple changes to natural ecosystems, to agriculture and forestry. Almost every aspect of the environment will be affected.

Much depends on how plants respond. With higher concentrations of carbon dioxide, plants grow faster, fix more nitrogen, use water and nutrients more efficiently. But not all plants thrive equally well. Plants such as maize, sorghum, millet, sugar and many pasture grasses, known as C4 plants, evolved during a cool period when carbon dioxide levels were low. They respond only modestly to higher levels of carbon dioxide. Most crops do better: in tests with plants grown under doubled carbon dioxide levels, grains had average yield increases of 36 per cent, fruit 21 per cent, leaf crops 19 per cent and pulses 17 per cent.[29]

But increased temperatures will bring problems too. The

number of hot days on which plants suffer heat stress will increase. Pests and diseases – until now killed off in temperate winters – will increase their range. Plant respiration will increase. Above 25°C this may cause yield losses which outweigh the gains from carbon dioxide.[30]

Global rainfall will increase. This too could mean increased plant growth overall. But the regional picture is likely to be very diverse.

The northern parts of Europe, of Russia and of North America are likely to gain. Average temperatures will increase more than the global average. The growing season will lengthen. The climatic limits of agriculture will shift northwards – though in some areas the soils will not be suitable.

In South and South East Asia summer rainfall and soil moisture may increase by 5–10 per cent. Rice production may increase as a result. The impact on the Sahel is uncertain. The equatorial rainbelts that bring seasonal rains might be expected to penetrate further inland. Against this, summer soil moisture may decrease marginally as plants respire more.[31]

But there will be losers too. In some equatorial regions rainfall is already so high that soils are excessively leached, and fertilizers are quickly washed away. Crops rot in the field and in storage more easily. As global temperatures rise, rainfall and humidity in these regions are likely to increase, and the problems with them.

Some regions will get dryer. Most climate models agree in showing central North America and Southern Europe suffering a significant drop in summer rainfall and soil moisture. Grave water shortages are likely, with severe competition between agriculture, industry and domestic needs. In these regions crop-yield potential may decrease by up to 30 per cent by the middle of the twenty-first century if carbon dioxide levels double. Western Arabia, the Maghreb, western West Africa, the Horn of Africa, and Southern Africa are also vulnerable. Coffee and tea production in many tropical countries will have to retreat up hill as temperatures warm up. In many places they may be squeezed out altogether.[32]

Almost everywhere there will be changes, and adaptation will be the order of the day. The geographical limits of crop boundaries will shift from one year to the next. In Europe the

northern boundary for growing maize will move northwards at
10–15 kilometres per year. The US corn belt could shift north-
eastwards by 175 kilometres for every 1°C of warming. Farmers
will have to adopt new crop varieties and new practices of water
management.[33]

Adjustment will be more painful and more costly in the
poorer developing countries. The higher the share of population
working in agriculture, the greater the numbers vulnerable to
damaging change. In 1989 three out of every five workers in the
labour force of developing countries depended on agriculture.[34]

The higher the share of national income made up by
agriculture, the more a country as a whole stands to lose if
agricultural production falls. In 1988 agriculture still made up
about a third of Gross Domestic Product in sub-Saharan Africa
and South Asia.[35]

During the transition period, when the centres of surplus
production are shifting, world food security will become shakier.
The North American corn belt is the major source of grain
traded on the world market. The hot dry year of 1988 gave a
foretaste of what may lie ahead. Drought cut coarse grain
production in North America by 30 per cent. Exports were kept
up only by drawing down stocks drastically. North American
cereal stocks plummeted from 223 million tonnes in 1987 to
only 72 million in 1990. As a result, world stocks dropped from
456 million tonnes to 301 million; from the safe level of fourteen
weeks' consumption, to just nine precarious weeks. If global
warming proceeds apace, years like 1988 will come along with
increasing frequency. Two or three in a row would push up
world grain prices, increase malnutrition in vulnerable groups,
and reduce stocks to danger levels.[36]

The multiplier effect

Global warming could multiply almost all the processes of
environmental degradation which this book has surveyed.

Land degradation could accelerate. With heavier rainfall,
erosion in tropical countries will increase. Higher temperatures
may worsen salinization in semi-arid climates, and spread
desertification further in North Africa and the Near East.[37]

Global warming could become an agent of massive deforestation. One estimate suggests that the natural range of forest will shrink from 58 per cent of global land area to only 47 per cent if carbon dioxide levels double. The potential range of tropical forest will expand – from 25 per cent to 40 per cent. But the forest will not be able to take advantage of this expanded range because much of it is already occupied by agriculture. The main shrinkage will come in the cold northern forests, which could decline from 23 per cent of the land area to only 1 per cent. In a warmer world forest fires, too, might increases in severity, as might the incidence of forest pests and diseases. Strong winds like those which destroyed millions of trees in Britain in 1988 may also become more common.[38]

Climatic boundaries may shift polewards by 30 to 36 kilometres every ten years. Most trees could not keep up with such a pace, and would be unable to move into their new natural ranges without human help. The loss of other species will accelerate. Many species would lag behind the shift in climatic zones. They would find themselves trapped in zones for which they were not suited, crowded out by exotic competitors better able to spread.

Areas dominated by human activity – cities, agricultural areas and pasture – will raise barriers that cannot be crossed. Isolated national parks, set up to conserve species, will become death traps. Few extend over the 300 to 450 kilometres that would be needed to allow adjustment to global warming of 3°C. Species living in the poleward ends of parks, and on the upper reaches of mountains, will be left with no retreat. The tundra zone could virtually disappear from Eurasia, along with all the species that dwell in it.[39]

Sea-level rise (see Chapter 16) compounds the threats to wetland habitats. They will be squeezed between rising seas on one side and farming and urban areas on the other. Any coastal defences that are built against rising seas will totally alter wetland ecology.

Global warming also poses a menace to coral reefs. Corals can grow upwards at between 1 and 10 millimetres each year. The best guess is that sea levels will rise at 6 millimetres per year. Some corals would be left behind at depths they are not adapted to, and would eventually die.[40]

The driving forces

Clearly human-induced atmospheric pollution and climate change are already causing enormous damage to the environment. They look set to cause much more. As always, population growth, consumption growth and technology change have worked together to increase the human impact.

Consumption has been important here. The spread of cars and of refrigeration has been a change in technology and consumption combined.

The role of technology is central. Fossil fuels have been an important source of energy since the seventeenth-century wood shortage. Chlorofluorocarbons came from nowhere in the 1930s. Methane production is inseparable from irrigated agriculture and livestock-rearing. Nitrogen oxides are a by-product of high temperature combustion.

But as ever, technology always works in conjunction with population.

Where technological change is rapid, the efect of population growth is relatively small. In the case of chlorofluorocarbons, an entirely new technology mushroomed from nowhere. Between 1950 and 1985 world population grew by 1·9 per cent a year – but CFC use grew by 9·1 per cent. Thus population growth accounted for at most 10 per cent of the overall increase. The other 90 per cent was due to changes in consumption and technology.[41]

In the case of sulphur dioxide emissions in Western countries, population growth played the role of counter-agent, reducing the effectiveness of controls. Per capita emissions in the OECD actually fell by 46 per cent between 1970 and 1988. But the population grew by 15·5 per cent, so that the overall reduction in emissions was only 38 per cent a year.[42]

Other pollutants are much more closely linked with long-established technologies, and with fundamental energy, food and transport needs which must be met for each member of the population. In these cases the impact of population growth is much stronger. Indeed in many cases it is the dominant factor, outweighing changes in consumption levels and technology put together (see Appendix, pp. 307–14).

Take carbon dioxide emissions. Over the 1960–1988 period

population growth accounts for 46 per cent of the increase in carbon dioxide emissions from fossil fuel burning in developing countries. In developed countries the share is 35 per cent. Deforestation is predominantly due to clearing of land for agriculture and urban settlements. If, as we suggested (p. 109) population growth was responsible for around 79 per cent of deforestation between 1973 and 1988, then it would have accounted – after allowing for forest thinning by logging – for perhaps 70 per cent of carbon dioxide output from land-use changes.[43]

With methane output the links are stronger still in developing countries where population growth is fastest. Irrigation is the main source of man-made methane. Over the 1961–88 period the population factor amounted to 82 per cent of the growth in irrigation in developing countries. In developed countries increases in irrigation were more a matter of improving profits than of meeting basic food and livelihood needs. Here the population impact was only 44 per cent.[44]

Livestock are the second largest artifical source of methane. In developing countries, between 1970 and 1989, the population impact on growth in livestock numbers was 69 per cent, and 59 per cent in developed countries. Technology contributed to a decline in livestock numbers for each unit of meat or milk production.[45]

Stages in management

The combination of population and consumption increases with technological change has increased the burden of air and atmosphere pollution. With gaseous wastes, we have entered the crisis phase after millennia of scattering, and a mere two hundred years of dumping. Serious controls have been in place, in developed countries, for a brief few decades. Sustainable atmosphere and climate management lie probably several decades in the future.

The first gaseous waste to be controlled was the most visible, most obviously linked to human health. Smoke had been a problem for three centuries in London. Yet it took the murderous London smog of 1952, which claimed 4000 victims, to bring about serious attempts to control air pollution.

The early responses did not control the problem at all, merely spread it and exported it. Factory chimneys were raised, so smoke was blown away from urban areas. Domestic solid fuel was controlled. People used electric fires instead. The pollution problem was shifted to big power stations sited away from cities.

Acid damage to freshwater fisheries and forests was next to become a cause for widespread concern. Acid rain was first named and analysed way back in 1872, when the UK's first pollution inspector, Robert Angus Smith, published a 600-page book on the subject. But it took a century before any action was taken. And it was taken on the easiest substance to control: sulphur dioxide. Once home coal fires are banned, this is emitted by relatively few large sources, which are easy to police. In the United States just 200 large coal-fired electricity plants were responsible for 57 per cent of sulphur dioxide emissions in 1986. Oxides of nitrogen – also important in acid rain production – have been left till later, since they involve expensive controls on emissions from hundreds of millions of private vehicles.[46]

Ozone came next in the control race, and for speed of response it beat all contenders. International reduction targets were agreed at Montreal in 1987, only two years after Farman's paper on the ozone hole was published. When it became clear that these targets were inadequate, more stringent controls were brought in within three years.

The swiftness of this response was encouraging. But the circumstances were unusual as environmental hazards go. This was no vague future menace, no slow and barely noticeable change. The ozone hole leapt fully formed into human awareness. It was huge from the moment it was noticed. There was a direct threat to human health, involving one of the most dreaded of all ailments – cancer. Every northern politician was personally threatened. No political pressure was needed to make the point.

And the solutions were not onerous. Chlorofluorocarbons were not essentials of life. Substitutes were already available.

The biggest problem of all – global warming – is quite

different. The impact is not yet visible, only predicted. It is rare indeed for serious environmental controls to be imposed before losses and costs are plain for all to see.

And carbon dioxide generation is central to our basic ways of life. Deforestation is the principal source of new cropland in the South. The burning of fossil fuels is the current basis of Western material civilization. Increase in domestic and industrial energy consumption per person is an indispensable element in raising incomes in developing countries. And the number of persons is growing fast.

The easiest substitute – nuclear energy – is politically unpopular, and rightly so. Alternative energy sources are still some way from cheap availability. Even if they expand rapidly, they must do so from a small base, so their absolute impact will be small at first.

Control of carbon dioxide emissions will test to destruction the human ability to respond to environmental challenge.

16

SORROWS COME NOT SINGLE SPIES:
Hatia Island, Bangladesh

The villages have all become rivers
And there is no safety for the land.
What can we use to check the waters?
Song of Hu-tzu, on the flooding of the
Yellow River, c. 109 BC

Hatia Island, in the Bay of Bengal, is not an easy place to get to.

You first take a battered, blacksmith-made bus on a bruising ride south of Noakhali in Bangladesh, to a landing that is no more than a slither of fine grey mud. There you board a shallow cattle boat and chug smokily alongside a shore where fish eagles duel over snakes caught in the undergrowth of mangrove forests. It is a rich habitat: terns, cormorants, herons, egrets, fish the waters. Curlews and whimbrels prod the beaches, sandpipers and herons plod along the glutinous mud.

The boat glides forward over waters cloudy with their burden of suspended silt. The shallows here change shape so fast that a chart drawn up one week would be out of date the next. A boatman in the prow sounds the bottom with a bamboo pole, holding up one finger, two fingers, three fingers, to indicate the clearance in cubits. At times you can hear the propeller grinding against the soft bottom.

The oppressor's wrong

Poor people, desperate to survive, fill every meagre niche. They hack down mangroves along the shoreline, out of sight of forestry officials, and sell them for firewood. They float on antique fishing boats shaped like Noah's ark, piled with wood for the blackened stoves on which they cook their meals. On glistening mudflats that barely top low tide, they seem to walk on water, tugging ugly, pink-bellied, razor-toothed mudfish out of the slime.

After an hour, the island appears on the southern horizon: a low hump, only 2 metres above sea level, 72 kilometres long, 13 broad at its widest. Like most of rural Bangladesh, it looks idyllic. Copses of palm, mango and bamboo surround the thatched houses. Each tiny hamlet has its own fish pond cum bathing and washing tank.

But beneath the surface lies a different reality. Life here is as harsh as in the dryest desert.

Unlike much of mainland Bangladesh, where two or three crops a year can be grown with irrigation, this land produces only one. It cannot be irrigated, because the groundwater is salt down to 100 metres. Much of the land is subject to saline floods, which damage crops and depress yields.

The population is around 300,000, the cropland only 40,000 hectares. The population density is 750 per square kilometre of cropland – almost twice as dense as the Netherlands. Dividing the land among the number of families gives an average holding of around 1 hectare. With yields of less than 2 tonnes of rice per hectare, this would just feed an average family of seven.[1]

But in unequal countries averages are always misleading. At the top of the Hatia pile are a small number of landowners who control far more land than they actually own on paper. They are called *jotdars* – literally 'force leaders'. Hatia has around twenty. They respect each other's property, but no one else's. Within his territory, the *jotdar* regards it as his hereditary right to grab all the land he can. Indeed he behaves as if it were his duty, and rarely misses an opportunity to extend his holdings. A common method is to lend money against collateral of land. When the borrower fails to repay, he forfeits his land, which is usually worth far more than the loan. Fraud and open violence are also common. Through marriage and patronage each *jotdar* controls a mass of allies, clients, and camp followers. Through bribery he often controls the police.

Such methods are everyday fare throughout Bangladesh.

Population growth gives them much of their impetus. It
weakens the poor, by continually fragmenting small holdings
below subsistence level and creating a reserve army of surplus
labour. But it also spurs the land-grabbers. They know that
when they die their holdings will be split among two or three
sons. 'Poverty forces us to be ruthless,' said one peasant. 'If we
are not clever and cunning we will have no chance of eating rice
every day and we will leave our children without land.'[2]

For the majority things stand far worse. Only one farmer in
five on Hatia owns and works his own land. Most of these are
smallholders. Some 44 per cent own a little land, and sharecrop
extra. The rest are pure sharecroppers. They give the landlord
half the crop at harvest. Many landlords demand extra favours
like free eggs or straw. The sharecropper is left with about 1
tonne per hectare – not enough to feed an average family.

There is no shortage of people willing to accept these
oppressive terms, for there is a pool of landless labourers –
about one family in six – who don't own or operate any land,
but work as labourers for others. Wages range from about 10
taka per day in winter in the south of the island, (about 15p or
30 US cents) to 35 at harvest time in the north. Rice costs
around 12 taka a kilo, so the best daily wage is not enough to
buy a day's rice for the average family.

Wives are worse off. They eat last, after the men have had
their fill. Most women are seriously malnourished. Since
purdah is strict, women are totally dependent on their menfolk.
Yet they have little security. Divorce is as easy as saying 'I
divorce thee' thrice. As sheer survival gets harder, many
husbands are unable to feed their families, and abandon them
to seek their own survival on the mainland. If they are lucky
they might remarry and pick up a second dowry.

Divorced and abandoned women and widows are the bottom
of this cruel heap. And there are many, for life here is short and
dangerous. They rarely remarry. They become breadwinners
for their children, but to do so they have to face the stigma of
working outside the home. They collect fuelwood for sale. They
husk and parboil paddy. They do domestic work for the better-
off, or grub the shorelines for mudfish.

If you see women working, chances are they will be widows
and divorcees. Beyond the southern embankment, the dry

season fields are pink and purple with reeking hilsa fish drying. Here groups of women and children pick the heads off tiger shrimps. Among them, a green shawl draped across her face from modesty, Zoar, grey-haired though not yet fifty, widowed sixteen years ago when her husband died of typhoid. She lives by begging, domestic work, and day work for the fishermen. For de-heading shrimps she gets paid 1 taka for 2 kilos. In a day she can manage 18 kilos – 9 taka per day – 13p or 26 cents.

Her eldest daughter, Feroja Katun, works beside her. Like all Bangladeshi women she is thin and gracile. She married a year ago. Her husband came to live with her and Zoar, and his earnings helped them to survive. But his parents were also desperately poor, and wanted his money to go to them. They persuaded him to divorce her. Feroja is still only twenty-two. A lifetime like this stretches ahead of her.

Fish eke out the poverty incomes. Every creek and every canal, every flooded paddy, every stretch of shore, is scoured for fish. But here too the sheer pressure of population has reduced the average take.

Fisherman Jamini Kumar Jalodas has a small wooden boat. He used to fish off the northern shore and would catch ten or twenty baskets on the average day. Now the catch there is only four or five. 'There are too many boats,' he complains. 'The nets are getting bigger, and the weave is finer. Now we have to sail forty miles south – two days' journey – to get a decent catch. But everyone else is doing the same. So catches there are getting smaller too. Sometimes we catch nothing at all. Only boats with motors can get far enough to be sure of a good catch.'

The rising seas

These tribulations are routine in Bangladesh, fifth poorest country in the world, with average incomes lower than Burkina Faso's.

Population growth and lack of other employment openings swelled the number of sea fishermen, from 200,000 in 1975 to almost half a million in 1988. The total catch at sea has continued to rise so far – but because of the intense competition the catch per boat and per day's work has been declining. Inland waters are seriously overfished, and the total catch is falling.[3]

Population growth has also slashed the average farm size from 1·45 hectares in the 1960s to only 0·9 hectares in the 1980s. Indeed in 1983–4 27 per cent of rural households had no land at all, and another 40 per cent had less than 1 acre (0·4 hectares). And as the number of farms below subsistence level grew, so did the numbers seeking work. Real daily wages fell by 46 per cent in the two decades up to 1983.[4]

As if life were not harsh enough, floods occur on an almost annual basis, and tropical cyclones curve over, sweeping in from the Bay of Bengal.

Global warming will increase the vulnerability of Hatia, and of Bangladesh as a whole, to floods and cyclones.

Current projections suggest that the average sea level will rise between 30 and 100 centimetres over the next hundred years, due to melting of glaciers and thermal expansion of the ocean as it warms.

Sea-level rise will be pure loss, with no compensating gains. The United States could lose 20,000 square kilometres of land valued at $650 billion. Britain could lose large areas of prime agricultural land in East Anglia. Venice is already flooded forty times a year, four times more than a century ago, thanks to a local sea-level rise of only 25 centimetres.[5]

Developing countries would be worst hit. They would be less able to absorb losses, less able to afford protection and adaptation. Thirty-four nations would have to spend more than 1 per cent of their gross national product to protect their coasts against the projected rise.[6]

Coral atoll nations are most at risk: in the Maldives, Tuvalu, Kiribati, the Marshall Islands, and Tokelau, half a million people live within three metres of sea level. But in terms of numbers of people affected, low-lying river deltas are the most threatened. They are among the most densely populated areas on earth. The Yangtze and Hwang-Ho deltas in China, the Irrawaddy, Mekong, Indus, and Niger deltas would be badly affected. A sea-level rise of only half a metre could flood prime farmland up to thirty kilometres south of Alexandria, an industrialized city of 3 million people. Egypt could lose 15 per cent of her cropland, and see 16 per cent of her population displaced.

When all impacts of global warming are taken into account, Bangladesh could lose more than any other sizeable nation on earth. If sea levels rose by one metre, some 2000 square kilometres – a sixth of the land area – would be lost. A tenth of the population – almost 12 million people in 1991 – would lose their homes, lands and livelihoods. They would be forced to move into inland areas that are already overpopulated.[7]

On remaining coastal land, sea-level rise would worsen the problem of storm surges. Floodwaters would be deeper and would penetrate further inland.

Rainwater floods would worsen too. Glaciers in the Himalayas would melt more rapidly. Rainfall over the subcontinent could well increase. The volume of water flowing downstream would swell, yet higher sea levels would make drainage slower.

These threats are not vague imponderables for Bangladesh. They are already regular occurences whose human and economic costs are known. In 1987 the Ganges recorded its highest flood peak ever. Two-fifths of the country was flooded. Thirty million people were affected. The damages totalled $0·5 billion – 2·5 per cent of the country's GNP. The following year's floods were worse. Record peaks on the Brahmaputra coincided with a high Ganges. Three-fifths of the country was flooded, affecting 45 million people. The damages of $1·5 billion amounted to 6 per cent of GNP.[8]

About five serious cyclones a year hit Bangladesh, mainly in the spring and autumn. There is no evidence that they have been increasing in frequency, but it is possible that they might in future. At present tropical storms develop only over seas warmer than 26°C. In future wider sea areas will reach this temperature.[9]

The dispossessed

The human consequences of global warming are hard for outsiders to imagine. But we do not have to rely on imagination alone. Land loss, flooding and cyclones are regular features of life on Hatia.

The island lies athwart the main outlet of the Lower Meghna river. Here the combined waters of three of Asia's mightiest rivers, the Meghna, Ganges and Brahmaputra, draining most of the Himalayas, finally reach the sea. Even in the dry season the current flows powerfully, sweeping boats along at 10 or 12 knots without sail or engine. In the monsoon the Meghna is like a fireman's hose pointed straight at Hatia's northern shore. Its massive load of silt gives it an erosive force as powerful as a sanding belt.

Land here has always been almost as fluid as the water that shapes it. The geography of the delta changes, like a speeded-up version of continental drift. Since 1789 – the date of the first reliable map of the area – Hatia has crept 50 kilometres southwards.

But in recent decades land loss has grown worse. During the sixties one of the Meghna's outlets – the Bamni river – was poldered to build up new farmland south of Noakhali. This speeded up the flow of the channel opposite Hatia, and the island's northern shores began losing farmland at a terrifying rate. Since 1970 some 68 square kilometres have been washed away.

River erosion has provided a harrowing dress rehearsal of the chaos and suffering the sea-level rise will cause. Since 1970 a quarter of the island's 40,000 families have lost their lands and homes. The impact has been harsh. About a third left the island, for Noakhali, the Chittagong Hill Tracts, or the slums of Dhaka.

The rest remained. Most headed for the long curving embankment built to protect the island to the south. Others live cheek by jowl in crowded colonies along the northern shore.

Three out of every four dispossessed families have seen their incomes fall by half or more. Before the erosion, three-quarters of them were farmers. After, only one in fourteen. More than half became day labourers doing seasonal agricultural work or casual porterage around the landings and bazaars.

Most of the victims have been disinherited many times as the river pushed them inland. Shujayet Hossein, fifty-year-old father of five sons and three daughters, was making his eleventh

move. Only the freestanding kitchen of his old house remained
among the banana palms, six metres from the flooding river.

He has never been a fortunate man. After his father died,
when he was a boy, his mother begged. Shujayet survived an
attack of smallpox, but was left with a permanent needling pain
in his feet. Now he can do no normal work, but earns a living of
sorts, making amulets for sick people.

He has learned to tolerate his sufferings from day to day by
not thinking about them. When he recounts them, he is on the
verge of tears. He squatted on the bare plinth of his old house,
on the brink, pointing five kilometres out to sea where his first
house stood.

The first move made the family landless. Each subsequent
move has thrust them deeper into poverty. Each time they have
to pay labourers to shift the props and woven palm walls and
rebuild them, usually with new thatch. The cost devours
whatever savings they have scraped together in the meantime.
Last time they borrowed 1000 taka (about £17 or $34) from a
bank. It was more than a month's wages, sliced out of a budget
already below the breadline. They couldn't keep up the
payments. The bank took them to court. Their cow, worth three
times the original loan, was seized.

For the present move they borrowed another 1000 taka. Since
they've no collateral left of any kind, the local moneylender
demanded interest payments of 200 taka a month – an annual
rate of 140 per cent. 'We have to work very hard to repay the
loan,' Shujayet's wife, Shakina, laments. 'Sometimes we starve
to repay it. If there's no work we don't eat.'

Three of their children are grown up, but have problems
surviving themselves and cannot help their parents. One son
has left for Noakhali. Another moved to the south of Hatia.
Twenty-two-year-old Rukiya, their eldest daughter, was abandoned by her husband while pregnant with their second child.
He kept complaining that he couldn't afford to feed his family.
One night a year ago he disappeared, and has not been heard of
since. Such abandonments are common in Hatia. The deepest
poverty can dissolve all moral bonds.

The Hossein's new site is a mere 70 metres inland, on the
protective mud embankment. Hundreds of families have built
their flimsy homes here, only a metre below the spine. They

know they will be washed away as soon as a flood overtops the wall. They know the river will advance again. They know they'll have to move again next year.

So why didn't they move further south, further inland, where they might gain a decade or more of respite from further erosion?

'We can't afford to transport our belongings down there,' Shujayet explains. 'We can't afford to buy land.'

Like sand martins, landless families line every embankment on the island – even on the seaward side, where they have no protection at all against storm floods. They live there because they have no choice. The embankment belongs to the government. It is the only land that is free. Corrupt officials and police exact illegal rents to allow them to stay. Desperate need forces them to comply.

If global warming proceeds on projected trends, Hatia's tragedies will be multiplied a hundredfold across Bangladesh as land is lost to sea-level rise.

The battle for Dhal Char

The river, as it slows, dumps soil washed from the northern shore to the south, as shallow grey mudflats, visible only at low tide. Gradually they rise above high-tide level, and become what locals call *Kak char*, crows' land, from the birds that hover searching for crabs. Salt-tolerant low grasses appear, then mangroves with curious aerial roots surrounding the trunk like a pincushion, then wild rice.

It may be another fifteen or twenty years before salt levels fall low enough for farming. But in Bangladesh population pressure drives people to occupy it sooner than they should, while it is still saline, subject to seawater floods at every high tide, endangering crops, livestock, houses, human lives. The production over the whole year from each hectare is only a half or two-thirds of what can be got on good land protected by the embankment.

If Hatia's population was static, those who lost land in the north could be compensated in the south. But because the population is growing, there is never enough new land, and the numbers of landless grow continually.

Competition for land in Bangladesh is murderous, often literally so. Those who need it most are least likely to get it. People dispossessed by erosion have no automatic right to newly accreted land. All newly deposited *char* land belongs to the government.

In theory it should be allotted to the landless. But there is a complicated rigmarole of form-filling at several offices. The landless don't know the ropes, and being illiterate can't fill in the forms. *Jotdars* and other rich landowners get hold of legal title to *char* land under fictitious names. Then they sublet the land to sharecroppers, and take half the harvest. A former MP for the island told me that three-quarters of all applications for government land were bogus.

Dhal Char is a featureless mound of infertile land exactly half way between Hatia and the island of Manpura to the west. It first emerged from the waters during the 1940s. As soon as it was cultivable, rival *jotdars* from the two islands laid claim to it. Every year they sent sharecroppers over to farm it. Every year, at harvest time, each one ordered his own tenants to seize the harvests of the opposing side. Old timers on Hatia remember the battles with fire in their eyes. Jobial Haq points proudly to the scar on his eyebrow where he was hit with a cudgel, then lifts his grimy vest to show where he was stabbed in the kidney.

After Bangladesh won independence from Pakistan in 1971, a young freedom fighter, Rafiq Alam, negotiated an agreement between the two *jotdars*. The island was divided between them. The *jotdars'* war for Dhal Char was over.

Fourteen years later the peacemaker himself launched another war. In 1985, with Oxfam funding, Rafiq Alam founded Dwip Unnayam Sangstha, the Island Development Society. DUS has since formed 550 groups of landless right across the island, and has helped hundreds of them to apply for government *char* land. In 1989 DUS got temporary leases for 457 farmers on Dhal Char. Azaher Uddin, the wily old *jotdar* who used to control Hatia's half of Dhal Char, let that year's harvest pass without taking action.

But towards the end of 1990, seeing his empire collapsing, he decided to make a stand. Whenever he met one of the Dhal

Char farmers, he shouted that if they didn't give him his rightful harvest share, he would take it from them forcibly.

'We met,' group leader Kamal remembers, 'and decided to defend ourselves. We built a shelter. Fifteen of us slept there through the harvest period to guard the crop. Most of the paddy was safely boated across to Hatia. Then we let our guard down.'

On October 15, at two in the morning, they were roused by a furious beating on the palm-leaf walls of the guards' hut.

'I woke up and found myself surrounded by Azaher Uddin's men beating me,' thirty-seven-year-old Mohiuddin remembers. 'They were shouting "You're the bastards who are helping the landless, you should get off this island." They said if I cried out they'd kill me. They kicked me, all over my body, and hit me with their sticks. When they'd finished I couldn't move. I had several bones broken. And they stole 800 taka (about £11) in cash.'

Johir Uddin, a quiet forty-year-old with greasy black hair, tried to escape. They ran after him. Then he felt a terrific blow on the head, and blacked out. He'd been hit on the skull with the blade of a home-made bamboo spear. When he woke up in the health centre on Hatia, two days later, he couldn't walk or see. It took him six weeks to recover. During that period he had to borrow 3000 taka (£42) to feed his family.

Living dangerously

Disaster – individual or collective – is never far away in Hatia. During the monsoon rains there are freshwater floods. For the rest of the year spring tides bring saline floods twice a month that ruin crops. And Hatia lies in the killing path of cyclones, averaging one or two each year.

Population pressure and destitution force the poor to live inside the open jaws of catastrophe. And those jaws have closed many times. Everyone I met had lived through repeated floods and cyclones. Many had lost husbands, wives, children, cattle, boats, furniture, homes.

Sayed Ahmed, leader of the landless in the southern *char*

lands, lost house and cattle in 1970. He saved himself by clinging to his thatch roof. His mother and father, who lived on the exposed, seaward side of the embankment, were washed away. Their bodies were never found.

Kamal, the Dhal Char leader, returned from the dead. He was swept out to sea clinging to a wooden linen chest. He hung on for twelve hours, battered by the waves, till the next high tide flung him ashore like a piece of flotsam.

I met fifty-five-year-old Arfuja Katun on the grey shore, bare feet shod in mud, tugging violently at the protruding roots of a mango tree, hoping to sell them as firewood. Her two-year-old grandson slept on the ledge above, head resting on a tree stump. She once had three hectares and ten cattle. The 1970 cyclone caught her husband on an exposed *char*, and swept him out to sea. In the following years her land was eroded away, and she sold off her cattle one by one. Now she keeps her six children by begging, domestic work, and gathering driftwood.

Along the northern shore a string of colonies for river-eroded families was established four years ago, behind a new embankment. Erosion victims of later years moved into the interstices. As many as a thousand families live in dense clusters of tiny thatch huts, within 200 metres of the sea. And the embankment was already breached.

The five colonies are socially segregated. Not wishing to rub shoulders with the unlettered, teachers, makers of sweetmeats, shopowners, poets, singers and actors, requested their own settlement. It is called Bodro, 'gentle'. 'A' 'B' and 'D' colonies are for farmers. The land 'A' was built on crumbled into the delta last year, and the inhabitants moved in with 'B'.

'C' colony is mainly for fisherfolk, avoided by others because of their malodorous occupation. A few months ago the embankment here collapsed. Jamini Kumar Jalodas and his family live three metres from the gaping breach. He once owned a yeomanly estate of seven hectares, eleven kilometres to the north. The 1970 cyclone initiated the family's long descent. They took refuge in high trees against the lashing winds and storm floods. A piece of flying tin roof hit their five-year-old boy on the head and killed him outright. Jamini still carries a scar on the hand with which he tried to protect his son.

In that storm the Kumars lost house, furniture, cattle –

everything except their land. Four years later, having painstak-
ingly recouped some of the losses, they watched helplessly as
the river swept their farm away piece by piece in the space of
four months. Since then they have moved house seven times.

Last year 'C' colony flooded on four separate occasions
through the broken embankment, losing furniture, boats, and
the fish stock in the pond.

I asked Jamini why he hadn't made his eighth move, since
the earliest monsoon rains would soon flood him out again.

'We should have moved last year,' he told me, 'but we don't
have the money. We're waiting for a hand from above. If the
government doesn't help, we'll have to move within the month.
The river is knocking at our door.'

The embankment protecting the houses of 'D' colony had
gone completely, and the residents were already moving.
Raman Hossein lost his house when the bank broke the
previous year. Four wooden beds worth £100 ($200) were
washed out to sea. He had just moved – for the fifteenth time –
only 50 metres inland, to the yard of his father-in-law's house.
He was hacking down his last mango tree, for sale to the
sawyers who sweat over huge hand saws in the main bazaar.

Poverty and population growth do not cause disasters – but
they intensify their impact. It was population growth on the
mainland that led people to move here, though they knew it was
prone to floods and cyclones. Growing population density has
worsened the toll of severe disasters in Bangladesh. 40,000 died
in the cyclone of 1822; in 1876, 100,000; 175,000 in 1897. In
1970 an estimated 300,000 people were killed, along with
280,000 cattle. 400,000 houses and 100,000 fishing boats were
destroyed.

And the poor are least able and least likely to protect or
insure themselves against disaster. Hatia island has some
cyclone shelters – cyclopean structures on concrete stilts that
double as village schools. A chain of Red Crescent volunteers
spread cyclone warnings by megaphone from house to house. In
1990 a warning was issued, but only a few hundred people came
to the shelters. 'The poor are least likely to come,' Red Crescent
officer Khairul Anam Khan told me. 'They can't afford to lose
their rice stores, their cows, their chickens. They stay at home
till the last.'

The island of silence

The most dangerous place on Hatia is Nijhum Dwip – the island of silence – just off the southern cape. It emerged as a salt flat in the 1940s, and was first settled in the 1960s. By 1969 there were 60 or 70 families, supplementing their meagre yields by fishing and gathering mangrove branches for sale as fuel.

Nijhum Dwip is still only a metre above sea level, and bears the full unsheltered force of cyclones. Those who came to live here knew they risked catastrophe in the long term. Survival in the short term gave them no option.

In 1970 the first settlers paid their debt to fortune. The 1970 cyclone, killed every living soul on Nijhum, save for one old woman, who clung desperately to a tree. Thereafter she became known as the tree woman.

Seeing a vacuum, *jotdar* Azaher Uddin leapt in to fill it. He laid claim to the land, and began to move new settlers in. In 1983 the *jotdar*'s rule was challenged. A co-operative applied for the land, and got a court order to evict Azaher Uddin. But the cooperative was a front for landowning middle peasants, some of them from the mainland.

DUS organized the landless sharecroppers on Nijhum Dwip to make their own application for the land. They refused to give Azaher Uddin his half share of the crop. His men burnt down the landless leader's house, and harvested crops in the dead of night.

But in 1989, with DUS help, 305 families of landless peasants got full title to the land, 0.8 hectares each.

Azaher Uddin left them in peace. But nature did not. On 29 April, 1991, only six weeks after I left Hatia, a powerful cyclone hit the Bangladesh coast and all the islands in the Bay of Bengal. Winds of 225 kilometres an hour coincided with high tide, sweeping before them a 7-metre-high wall of water. Some 440,000 cattle were lost, and 860,000 homes destroyed. The official death toll was 130,000, but this included only the bodies that were found. The true toll was probably 200–300,000.[10]

On Hatia 90 per cent of the houses were destroyed. The spring rice crop was lost. Deep tubewells were contaminated with salt water and sewage. Freshwater fish in house ponds were killed. Fishermen's boats were carried away and smashed to pieces.

The fields were littered with the floating corpses of people and animals. Four thousand bodies were counted, not including those washed out to sea. The worst casualties were in the south, where people were living outside the embankment.

On Nijhum Dwip half the population of 2500 died.

It will be a number of years before new settlers forget the trauma and step in to the empty lands there. But step in they surely will.

Last to rise out of the sea, the island of silence will be first to disappear again if global warming raises sea levels by as little as half a metre.

17

TOWARDS A GENERAL THEORY

The beginning of disaster is not much, as when a fire
burns small in its first stages and ends in catastrophe. As fire's
course is, such is the course taken by human misdeeds.
Solon, *Prayer to the Muses*, early sixth century BC[1]

We began by contrasting the conflicting views on population
and the environment. We end with an attempt at peacemaking.

All schools of thought are partly right. There is good evidence
to support all the chief contenders. That is why the controversy
has kept going so heatedly for so long.

The Malthusians are right when they focus on population as
a central factor. The left-of-centre radicals are right when they
highlight poverty, inequality, overconsumption by the North, or
technology. The right-wing radicals are right when they stress
the capacity of human ingenuity to adapt and the need for free
markets to enable it to do so.

Where these partial schools err is precisely in their partiality.
Whenever they elevate their own central concern into the
universally dominant factor, and downplay the others as
secondary or insignificant, they go wrong.

The real world is inconveniently complicated. It is a vast
system in which the immense complexity of human society
interacts with the even greater complexity of the natural world.
The character of that interaction varies from one place to
another. And in all places, it changes over time. We cannot
hope to encompass this complexity in a single handy slogan. If
we try to boil it all down to one or two factors, we will fail to
understand reality. If we act on only one or two factors, we will
fail to shift reality.

Nevertheless we must simplify to some extent, or we will be
unable to understand or to act. We need to move towards a
synthesis.

Demographers have a useful concept when investigating what determines the level of female fertility. There are a set of immediate factors – marriage, use of contraception and abortion, breast-feeding, and so on. These 'proximate determinants' are the only factors that directly determine fertility. Then there is that vast range of social, economic and cultural factors which work on the proximate determinants and influence fertility indirectly.[2]

It is useful to consider our familiar three factors – population, consumption levels and technology – as the proximate, direct determinants of environmental impact. The three factors influence each other. Many other factors influence each of these, and are influenced by them in turn. We shall look at these indirect factors in the next chapter.

But let's begin with our three protagonists.

The consumption factor measures how many goods each person uses and how much waste they create. Technology determines how much in the way of resources are needed to produce each unit of goods, and how much waste is emitted into the environment after consumption. Population is simply the number of persons: the multiplier needed to arrive at total consumption or output.

The population and consumption factors are relatively straightforward. Whenever they increase, other things remaining equal, environmental impact increases. When they decrease, impact decreases. Technology is double-edged. An increase in the technical armoury sometimes increases environmental impact, sometimes decreases it. When throwaway cans replaced reusable bottles, technology change increased environmental impact. When fuel efficiency in cars was increased, impact was reduced.

Often technological change does both. The coal-burning steam engine reduced deforestation: but kick-started the era of acid rain and global warming. Chemical fertilizers reduced the area of land needed to grow a given amount of food – but increased pollution of waterways.

Calculating the blame

All three factors are involved in almost every situation. Their relative contribution to environmental degradation varies considerably from one field to another, and even from one time period to another.

The seminal work in this field was done by US environmentalist Barry Commoner. Commoner lacked systematic data on actual ecological damage. So he measured environmental impact by proxy, through total emission of specific pollutants. The formula he used was as follows:

$$\text{pollutant} = \overset{1}{\text{population}} \times \overset{2}{\frac{\text{good}}{\text{population}}} \times \overset{3}{\frac{\text{pollutant}}{\text{good}}}$$

Item two measures consumption. It refers to any economic good that people directly consume, from food to vehicle miles. Item three is the technology element. It considers the amount of pollution emitted for each unit of the good.

Commoner applied the formula to a range of pollutants in the United States, including nitrogenous fertilizer. He found that the amount of nitrogen applied per tonne of crops produced increased by 9 per cent a year between 1949 and 1968. But population grew by only 1.6 per cent, and food consumption per person by a mere 0.6 per cent. Population growth therefore accounted for only 14 per cent of the combined increase, and consumption for a mere 5 per cent, while the change in technology was responsible for 81 per cent of the impact.[3]

Over a whole range of pollutants he found a roughly similar pattern for periods between 1946–50 to 1967–8. Population growth accounted for only 17–21 per cent of the increase in synthetic organic pesticides, and of nitrogen oxides and tetraethyl lead from vehicles. It was responsible for only 13 per cent of the increase in non-returnable beer bottles and a mere 10 per cent in the case of phosphorus from detergents.

Increases in consumption were even less significant. Growth in food consumption per person accounted for only 4–5 per cent of the increase in fertilizer nitrogen and pesticides. There was no increase in per capita consumption of cleaners to account for phosphorus emissions, and only a 0.3 per cent growth in beer consumption per head. Only in the case of nitrogen oxides and lead was growth in consumption significant, making up 35 and

42 per cent of the increase respectively. The reason for the latter was that vehicle miles travelled per person doubled between 1946 and 1967.

Much more recently, Commoner looked at pollution from nitrates, cars and electricity in 65 developing countries. Except for fertilizers, the data was much shakier and the methodology weaker. In the case of cars Commoner used gross domestic product per person as the consumption factor, instead of vehicle miles per person. For pollution from cars he used the number of cars as proxy. For electricity, GNP per person and consumption of commercial electricity were used. It could be argued that here he was not measuring pollution at all, but the increase in car ownership and use of commercial energy in relation to economic and population growth.

In the developing countries, population growth was faster than in the United States and accounted for a bigger share of changes: between 24 and 31 per cent of the increase in pollution. Technology was the dominant factor in all three cases.[4]

For the periods and pollutants Commoner is dealing with, technology change is certainly the dominant feature. 'Environmental quality,' he concludes, 'is largely governed, not by population growth, but by the nature of the technologies of production.'

But such a sweeping statement is not justified. Even his own results from developing countries, show that population was a significant factor. It accounted for between a quarter and third of the changes – well worth working on as a way of reducing pollution.

And his conclusions are heavily influenced by the choice of examples. The period 1946–68 in the USA was one of rapid technological change, while population growth was moderate, averaging 1·6 per cent a year. In several of the major areas he considers – food and beer, household cleaners, fibres for clothing – consumption per person had already reached saturation level. The developing country examples are also cases where rapid technological change is under way: the shift from organic to chemical fertilizers; the rise in car ownership from very low levels; the change from home-produced bio-mass energy to commercial electricity.

Measuring population impact

But the effort to measure population's share of the 'blame' for environmental damage is important. Concrete evidence and hard data help to get things into clearer perspective.

Since Commoner's examples are nearly all concerned with waste output of one kind or another, I examined a wider range of areas, using his basic approach with a somewhat modified formula:

environmental impact =
population × consumption per person × impact per unit of consumption

Impact covers any one of our three main forms of interaction with the environment: our use of primary resources from minerals to water and land; our physical occupation of space; or our output of pollutants. The formula makes it clear that, for any given level of consumption per person or technology, the more people there are, the higher is consumption of resources, the more space is occupied, and the more waste is produced. Slower population growth means slower increase in consumption of resources and space, and of waste output, other things being equal. Fewer people mean lower total consumption and waste. Of course, the formula also shows that higher consumption per person – holding technology and population constant – means greater impact. And that with population and consumption constant, technology changes also affect impact.

To assess what contribution population makes to increases in environmental damage – to try to assign relative blame – we have to look at changes over time. What we need is some measure of the share of each of our three elements in overall impact. If all three are pushing upwards, or, as may happen more frequently in future, downwards, we can simply express the change in each one in turn as a percentage of the total change. So here:

Population impact (as % of total change) = $\dfrac{\% \text{ change in population} \times 100}{\% \text{ change in use of resource or output of pollutant}}$

Where not all three factors are pushing in the same direction, it is useful to distinguish upward pressures from downward ones. So we can assess the share of the upward pressures out of +100 per cent, and the downward ones out of −100 per cent. The methods and results are given in detail in the appendix (see pp. 306–314). The results vary very widely depending on what you are looking at and when. Table 1 (p. 243) summarizes the findings – the first column shows the basic change in the area studied, the next three show the relative impacts of population, consumption and technology.

Population growth bears a relatively small share of the blame in cases of rapid technological change. Chlorofluorocarbon use exploded from zero in the 1930s, when CFCs first began to be used commercially. As we have seen (p. 217) population growth could account for at most 10 per cent of the increase in emissions between 1950 and 1985.

Technological change was, on the face of it, the principle factor in increasing use of fertilizers. It accounted for 62 per cent of increased fertilizer use in developed countries and 70 per cent in developing. Population growth accounted for 21 and 22 per cent respectively. Increase in consumption per person was responsible for the residue: Of course it could be argued that the technological change here was simply a substitute for land expansion, and was driven by population growth and increasing consumption.

In developed countries, recent years have seen a heartening new trend: in some fields environmental impact is decreasing. These include the major urban air pollutants. Emissions of sulphur dioxide in the OECD fell by 38 per cent between 1970 and 1988. This drop was entirely due to technological changes – about a third to increase in energy efficiency, and two thirds to flue scrubbing, changes in fuel mix and so on. Population and consumption continued to grow, making the results worse than they would otherwise have been. Population accounted for 25 per cent of the upward pressure and increased consumption for 75 per cent.

But there are other cases where population growth has been the principle factor. These usually involve very basic needs for energy, food and land, where the underlying technologies are changing more slowly. Here population growth usually outweighs the other two factors combined in developing countries. In developed countries, population growth is slower, and other factors such as commercial or technological change are stronger. Hence the population contribution is often smaller, but still usually significant.

Population growth is the dominant factor by far in deforestation and the resultant loss of species. As we saw in Chapter 7 (pp. 108–9), it was probably responsible for around four-fifths of deforestation between 1973 and 1988.

Irrigation contributes to salinization, waterlogging and increased methane output. Population growth outweighs all other factors in growth of irrigation. If we assume the same increase in consumption or *all* farmland (see appendix p. 308) then population growth accounted for 72 per cent of the increase in irrigated area in developing countries between 1961 and 1988. Livestock also emit methane and contribute to soil erosion through overgrazing. Population growth accounts for 69 per cent of the growth in livestock numbers in developing countries. In developed countries, where populations were growing more slowly, the population factor accounted for only 59 per cent of the change in cattle numbers and 46 per cent for irrigation, but it was still a very significant factor.

Population growth appears to be a significant factor in carbon dioxide emissions. It accounted for 46 per cent of the increase in fossil emissions in developing countries, between 1960 and 1988, and 35 per cent of the increase in developed countries. The weighted global average was 44 per cent.

So far we have been studying only the total level of resource consumption, or of waste ouput. The actual impact on the environment is determined by two further factors. One is the density of the activity – the degree of concentration. This often correlates closely with population density. For example, the excreta of a thousand people scattered over a wide area of land has little ecological impact save on human health. The same amount channelled through a single sewer outlet into a small river can alter the river ecosystem radically.

Table 1 SUMMARY OF POPULATION IMPACTS

	Overall % change	% share of change due to: Population	% share of change due to: Consumption	% share of change due to: Technology
Arable land growth 1961–85:				
(Contributes to deforestation, loss of wetlands, species loss)				
Developing countries	15%	+72%	+28%	−100%
Developed countries	3%	+46%	+54%	−100%
Growth in livestock numbers (cattle, sheep and goats) 1961–85:				
(Contributes to erosion, deforestation, methane emissions)				
Developing countries	36%	+69%	+31%	−100%
Developed countries	8%	+59%	+41%	−100%
Growth in fertilizer use 1961–88:				
(Contributes to water pollution, global warming, micronutrient depletion)				
Developing countries	1568%	+22%	+8%	+70%
Developed countries	208%	+21%	+18%	+61%
Change in air pollutants in OECD countries, 1970–88:				
(Various)		+25%	+75%	−100%
[Short method: population element only]				
Growth in carbon dioxide emissions from fossil fuels and cement 1960–88:				
(contributes to greenhouse effect)				
Developing	289%	+46%		
Developed	95%	+35%		

A minus score means that the factor was a downward pressure on environmental impact. See appendix pp. 306–314 for methods and sources.

We can talk of resource demand density in the case of resource use, and pollution density with waste output, or more generally of impact density.

The other factor is the carrying capacity of the environment. This decides the amount of damage actually suffered. We can define it as the level of resource use or waste output that can be sustained indefinitely, without long-term deterioration in the resource base.

Carrying capacity is a controversial concept. It has two aspects: the productive carrying capacity – the ability to provide resources such as food or minerals. And the waste carrying capacity – the ability to absorb a certain level of pollution or degradation without significant damage. For example, if soil erosion does not overstep the bounds of soil formation, no harm is done. Carrying capacity is rarely an exact term we usually only discover its limits after we have exceeded them.

In the case of resource use, carrying capacity is not static. As their numbers grow, humans intervene to raise productive carrying capacity. By sowing grain instead of reaping wild cereals, people raised the carrying capacity of the land. Most changes in agricultural technology are designed to raise carrying capacity. This flexibility helps to explain why resource use has so far managed to keep rough pace with population growth in most places.

The environment's waste carrying capacity seems to be far less flexible. The limits have already been outstripped in many cases – coastal pollution, damage to the ozone layer, acid rain, and so on. Paradoxically, many attempts to increase productive carrying capacity – such as fossil fuel and chemical fertilizer use – have led to waste carrying capacity being exceeded.

Technology might be able to increase waste carrying capacity. We can lime acid soils and lakes. We might one day fertilize the open seas to increase uptake of CO_2 by plankton. We might breed bacteria that can eat up plastics and oil. But playing around with ecosystems whose linkages we don't fully understand has often brought further damage. Frankenstein sincerely believed that he was contributing to the welfare of humanity.

From gatherers to managers

The three proximate factors are not independent of each other. Rises in population density, or in consumption, lead, through pressure on carrying capacity, to changes in technology. Changes in consumption may lead to changes in population growth, and vice versa. These connections have driven the major changes in human relations with the environment.

Boserup and others have claimed that in agriculture and industry, changes in technology are driven by increases in population density (see pp. 28–36). We can go much further than this and suggest a general theory of population-driven changes in resource management. In all aspects of human interaction with the environment, most major changes in resource and waste management are driven by changes in impact density – the combined pressure of our three elements of population, technology and consumption.

As population density and consumption have increased, so have resource demand and pollution densities. Whenever these passed beyond the carrying capacity of the environment problems have arisen. The search for solutions to these problems has led to changes in productive technology and in the way people manage resources and waste output.

We have seen similar processes at work in every sphere, from wildlife and forests, through farming and soils, to cities and wastes. And a similar sequence of phases (see Table 2, p. 248).

Let us consider the environment first as a bank of resources. In the beginning is abundance. Resources are vast in comparison with human numbers. Population density is low. There is plenty for everyone: all people need to do is to collect what is there. This is the gathering phase.

Gradually numbers multiply and exert increasing pressure on resources. Yet at first people continue with the technologies and attitudes of the gathering phase. The resource is depleted. This is the mining phase. In some cases, when fuelled by windfall profits, as in tropical forests, it reaches a paroxysm in pillaging.

Mining eventually leads to crisis. Potentially renewable living resources are not renewed, but depleted below the level at which they can renew themselves. Scarcities develop – hunted species disappear one by one, soil yields fall, firewood and water have to be sought further and further away. Depletion reaches the level at which it becomes visible, even painful.

At varying points during the crisis phase, adjustment begins. At first it is experimental – a few Cassandras, a few bold innovators, ignored by the majority. But as methods are found and prove effective, they spread widely.

Finally we reach the phase of sustainable resource management. Renewable resources are managed so as to produce a sustained output, whether of game, wood or food.

The transition to sustainability is not a clear, irreversible shift. It has to be continually re-won from setbacks. European farming in the eighteenth century, mixing arable and livestock with crop rotation, achieved higher yields than ever before in a sustainable way. Chinese farming reached this stage even earlier, in the fourteenth century. In the nineteenth century, chemical fertilizers raised yields dramatically – but led to pollution problems. Now we are in a second transition phase in which organic methods are returning to favour to complement much lower levels of chemicals.

A similar sequence of management phases applies to the environment as a sink for wastes.

Land, water and air are waste sinks with limited space or absorptive capacity. When sink space is abundant and population density is low, waste is disposed of by scattering on waste commons. Since wastes are mainly recyclable, this does no lasting damage. Controls exist, if at all, only for human excreta, to avoid transmission of disease.

As the size of settlements grows and population density rises, so does pollution density. Wastes are concentrated, but the scattering habit continues, so they are dumped – largely without controls. As industry and technology advance, the proportion of durable and toxic wastes increases. Yet at the same time sink space is shrinking.

The crisis with waste can be called the spoil phase. Waste output grows beyond absorptive capacity. In some cases humans are direct victims, and the response is more rapid – depending on the degree of democracy. In others it is renewable living resources that are poisoned and further depleted. In these cases adjustment takes longer.

As problems mount, there is a gradual move into the phase of sustainable waste management – reducing output, controlling emissions, recycling, and so on.

The resource and waste management phases move to some extent in parallel: gathering goes with scattering; mining goes with dumping; then come crisis and transition; and sustainable management.

Shifts in management go hand in hand with changes in ownership and control of resources. During the phase of abundance there is no need to control resource use or waste disposal. So there is open access. When a resource begins to be limited, it is taken into common ownership and control. When it becomes scarce it tends to be privatized.

The fluid sinks, air and water, cannot be privatized. So they are subject to regulation and control on an ever-widening scale, rising from local, to national, to regional, to international.

Each phase brings increasing requirements for political organization. When land is abundant, the community regulates only sharing of the proceeds and defence against other groups. As land grows scarcer, institutions such as chiefdoms are needed to allocate it. In the final stage, use becomes permanent and hereditary. The state evolves to guarantee property rights against theft and invasion, and to resolve disputes. The control of waste sinks requires a growing degree of regional and eventually global government.

Tables 2 and 3 show the sequence of phases related to various aspects of resources and sinks. To tie in with Boserup's sequence, I have added the extra phase of fallowing.

Followers of the cornucopia theory should feel themselves at home so far. For clearly they are right in their claim that population growth is a major player in driving technological and social change. They are also right when they observe that there is usually adaptation and, in one way or another, the problems of scarcity and pollution are eventually addressed and usually dealt with. Population growth helps to force the changes needed to cope with population growth.

The problems of adaptation

Where the cornucopians go wrong is in their complacency. All's for the best in the best of all possible worlds: so Dr Pangloss taught Voltaire's naïve hero, Candide, as he staggered from one catastrophe to another.

The human race, too, has tottered from one environmental disaster to another. In the 1970s fuelwood shortage, erosion, desertification, salinization. In the 1980s an extra chain of major emergencies on a rising scale: from red tides and forest death, to the ozone hole and global warming.

Adaptation occurs, indeed. But problems are what drives the major adaptations. And a problem that is serious enough to warrant a major change in technology is a big problem. Before the problems are resolved, there is often suffering or loss on a

Table 2 STAGES IN RESOURCE MANAGEMENT

Production system and resource state	Ownership	Plantfood	Animal	Fish	Wood
Gathering Abundant	*Communal*	*Gathering*	*Hunting*	*Fishing*	*Gathering*
Fallow- Plentiful	Communal	Fallowing burn, hoe	Pastoral, resting	Fishing, closed seasons	From cut fallow
Mining- Depleting	Mixed	Declining fallow	Over- grazing	Overfishing	Deforesta- tion
Crisis	Mixed	Falling yields erosion	Desertifica- tion	Stock decline	Wood shortage
Sustainable management	Private plus collective control of commons	Fertilizer, ploughing, irrigation, conserva- tion	Arable– livestock, controlled grazing	Controlled fishing, fish- farming	Forest manage- ment, agroforestry

Table 3 STAGES IN WASTE MANAGEMENT

Waste system	State of waste sink	Controls
Scattering	Healthy	None
Dumping	Pollution increasing	None
Spoil [crisis]	Poisoned, resource base or human health affected	Progressive, from local, to regional, to international
Sustainable management	Stable	Full range of controls. Global government

major scale. It is the suffering or the loss that drives the search
for solutions.

In our use of the environment, especially as a waste sink, the
adaptation has rarely been smooth. Often very severe damage is
done before adaptation comes about. In some cases adaptation
is delayed even when very high levels of damage have been
reached, where social and political institutions make adaptation
impossible. In many cases the damage can be reversed only at
high cost. In some it is irreversible for all time. Barring the
development of some DNA resurrection technique, we will
never again see a live elephant bird or passenger pigeon.

Even when a successful transition has been made to the
sustainable management phase, the final outcome is not always
an improvement on the state of affairs under the very first,
gathering phase.

Permanent farming, for example, involves far more labour
than hunting and gathering or shifting cultivation. Controlled
fishing, introduced after the stock has been depleted, will be at
lower levels than before. Waste recycling involves extra labour
for the household, extra kitchen and street space for sorting
bins.

And when population growth rates are rapid, it is a lot
harder for adaptation to keep pace.

There's many a slip

The crisis phase is pivotal. Its duration determines the extent of
damage done before successful adaptation occurs. It can be long
or short. In environmental matters it is often long.

There are four main stages: perception of the problem;
understanding of causes; development of technologies; and
widespread application of those technologies.

Delays crop up at every phase.

Perception seems simple enough: but in environmental
matters it is rarely so. It depends partly on the speed and
visibility of the change, partly on preconceptions which obscure
things even when they are staring you in the face.

Prophets and precursors may recognize early-warning signals. But for most people the threshold at which environmental

problems are perceived as problems is often quite high. Things
have to get serious before they're taken seriously.

That is not the result of human obtuseness. Environmental
impacts tend to build up slowly, often in a way that can't easily
be spotted. The threatened demise of a big cute animal like the
panda arouses great interest and attention. The massacre of
hundreds of species of insects per year in the rainforest passes
without even being recorded. The insects notice it. Humans
don't. Sheet erosion – where the whole field loses millimetres or
centimetres of soil a year – often goes unnoticed by farmers.

There are intriguing parallels with human health, where
problems also build up slowly and invisibly. Cancer begins in a
small way. By the time symptoms become visible, things have
usually reached an advanced stage where drastic action is
required, and cure is less likely.

Perceiving a problem as a problem is only the first step. The
causes have to be be understood before it can be addressed.
Muslims in Burkina Faso attribute desertification to punish-
ment by Allah of the sins of Westernized youth. The symptoms
of forest death were in Europe first perceived in the late 1960s.
The full range of interacting causes is still not fully understood
at the beginning of the 1990s. The ozone hole was spotted
before the complex chemistry of its development was worked
out.

Even when problems are perceived and their causes understood, the technology and social organization to deal with them
have to be developed. The length of time taken depends on the
extent to which existing technology and organization can be
adapted - and the complexity of the task to be dealt with.

Sometimes technology is pre-adapted. Life's transition from
water to land was eased because fins for swimming could be
modified, without too much difficulty, into legs for walking. Pre-
adaptation helps in the evolution of technology too. A technique
developed to solve one problem can be adapted to problems
that arise subsequently. Things go smoothest when change can
proceed by minor modifications of the technology that already
existed. Ploughshares cut deeper and deeper. Three-course,
then four-course crop rotations were introduced. Irrigated rice
production proved itself capable of infinite fine tunings to boost
performance.

But the course of adaptation does not always run smoothly. Blockages can arise. Most of sub-Saharan Africa is stuck in a technology cul-de-sac right now. There are too many elements that need changing, all at once. Africa needs improved hybrid seeds – few of which have yet been identified. She needs new fertilizer formulae. New ways of managing livestock. New ways of conserving soils and trees.

The parts of ecosystems are interdependent in complex ways. As yet our knowledge of these links is relatively poor. Hence today's miracle solution often turns into tomorrow's problem.

Technologies devised to solve a known problem often give rise to other problems that were not foreseen. Drugs are thoroughly tested before they are launched on humans, yet they still have unforeseen side effects. Most chemicals are not tested at all for their impact on the environment, where they will all eventually end up.

Fossil fuels, dams and nuclear power deal with shortages of energy sources – yet give rise to a whole range of other environmental problems. Several CFC substitutes, promoted to slow ozone depletion, will accelerate global warming.

Modern 'solutions' to population pressure on land cause pollution problems which will require further adaptation. Chemical fertilizers cause eutrophication. In drinking water they cause cancer. Pesticides damage wildlife. They remove natural predators and promote population explosions in pests as soon as they become resistant.

Nor is the problem solved simply by some clever invention. A technology will only be widely adopted if it is acceptable to most users. It must work socially as well as technically. It must fit in with users' economic and social circumstances. Many technologies that experts thought would solve Africa's problems have been rejected by farmers. 'Miracle' seeds that did well on good soils with plenty of water and fertilizer were no use to farmers with problem soils, little water, and no cash.

A further problem with technologies to preserve the environment is that they are often not marketable goods that can be bought and sold, like spades or seeds. If they were, market discipline would weed out inappropriate solutions. But environmental technologies are often general ways of doing things or managing things. You cannot buy or sell a method of terracing, a way of arranging shrubs along the contour, an organization for running a common forest, or a set of controls on fish catches.

Risking the earth

Another factor stands between perception of environmental problems and action: readiness to act. Without it nothing will be done.

Here are more strong parallels with health behaviour. When a clear and painful condition has arisen, most people ring the doctor immediately. Cures are in great demand. Prevention is far less popular.

But many conditions build up slowly and invisibly. You don't know if they are developing or, if so, how serious they might become. Prudence dictates the precautionary principle: it's better to take preventive action that run risks of mishap. Yet in real life a very large number of people prefer to take risks and continue with agreeable but dangerous behaviour, rather than take action to prevent something that may never happen.

It's not only with the environment that people take foolish risks: they regularly gamble with their own health and safety and those of other people. Some clients pay extra to have unprotected sex with prostitutes – though they may catch or spread HIV. The links between heart and lung disease and smoking are proven beyond reasonable dispute. Yet many people go on smoking. It often takes a first heart attack to induce people to change their diet and life-style. For many people that first attack may end their chances of changing.

Safety behaviour is similar. People who are not of a nervous disposition often need a break-in or a fire before they fit locks or fire alarms. A major loss of life, an expensive law suit, or both, are almost always required before private or public enterprises take costly precautions.

Risks, even those we know of, often have to become reality to be seriously confronted.

The parallel with global climate change is uncomfortably close. We know that the ozone layer is thinning: yet world leaders in 1990 agreed to a protocol that allowed a 50 per cent increase in CFC output before phasing out. We know that increased greenhouse gases may precipitate global warming. But we do not know for certain, nor will we know for thirty or forty years. We will not know for certain what the consequences are, unless we let them happen. And many vested interests argue that this is precisely what we should do.

The politics of environmental response

Readiness to act does not necessarily follow perception of the problem. Politics intervenes.

Readiness to act can be swift where a strong impact shows up on the polluter or depleter's own property. But frequently the costs of pollution or depletion are external. They fall on other people, other generations, other species. When this happens the channels through which problems become known and acted on lengthen. Conflicts of interest often mean that they are not acted on at all.

External impacts can only be resolved politically or legally. The outcome depends on the relative strengths of polluter and polluted. And the tug of war often favours the polluter.

There are differences in incentives to act. The polluter derives large benefits. Most victims of pollution suffer minor losses. Suppose, the Italian sociologist Pareto asks us, that there is a proposal to get 30 million people to pay out 1 franc a year, and distribute the proceeds to just 30 people. The 30 will move heaven and earth to get the proposal enacted. But few of the 30 million would shift a finger to prevent them.[5]

Polluters are often concentrated – factory owners and employees, for example. Pollution victims are scattered and hard to organize. When they are dumb animals or unborn future generations, they cannot speak for themselves at all but have to rely on champions.

Victims are often politically weak in other ways. Many are poor, since land prices and rents are cheap in marginal areas and polluted slums. In sensitive hill areas or rainforests they are often tribal people who do not speak the national language and have poor links with the political system.

Ranged against them are polluters or plunderers with resources to influence politicians and officials, to avoid regulation or enforcement. Pollution and pillaging pay. Excess profits accrue when a resource is plundered or polluted. The external costs are borne by society, nature or future generations.

18

PARTICULAR FAULTS:
sharing the blame

Few things in life have a simple cause. If a pedestrian is knocked down in the road, the immediate cause is the impact of a fast-moving lump of heavy metal on a soft human body. But no one trying to explain the event would stop there. Perhaps the pedestrian was daydreaming. Perhaps the driver was distracted by a pretty woman on the sidewalk. The view may have been obscured by an illegally parked lorry. The chain of causality extends in all directions.

And so it is with environmental damage. Population, technology and consumption impinge directly on the environment. But many other factors affect these three. By incorporating them in a holistic analysis, we can include almost all the elements that the various schools of thought consider important.

A couch for luxury

For the left, inequality is the fountainhead of all problems.

Inequality is multi-dimensional. It prevails in varying degrees between classes and sexes in every political and economic system. And it exists geographically, between regions of the same country, between countries, and between regions of the world. At every level it has some impact on the environment.

The rich–poor contrast is most heavily emphasized. It has become a cliché of environmental writing that two groups at opposite poles pose the greatest threats to the environment: the richest people on earth, and the poorest.

After three development decades, we still live in a grossly

unequal world. Indeed the share-out of the world's Gross Domestic Product has grown more unequal. In 1965 high income countries enjoyed 70 per cent of global GDP, while developing countries took 19 per cent. In 1989 the 16 per cent of the world's population who lived in rich countries garnered 73 per cent of the world's GDP, while the 78 per cent who lived in developing countries earned less than 16 per cent. Average incomes per person in high income countries were 23 times higher than in developing countries.

The rich are certainly a threat. The environmental damage that individuals create increases with wealth. Higher incomes mean higher consumption of resources of all kinds, and therefore higher levels of waste.

Northern countries are far and away the biggest consumers of non-renewable resources. In the late 1980s they were using 58 per cent of the world's fertilizer production, 75 per cent of the oil, 86 per cent of the natural gas and 93 per cent of nuclear energy.[1]

Western countries have been cleaning up their own houses in response to rising public awareness. But they are still sweeping a great deal of filth out of the back door, into the oceans and atmosphere. In the late 1980s, the developed countries generated 91 per cent of the world's industrial waste, 93 per cent of industrial effluents and 95 per cent of hazardous waste. They were responsible for 87 per cent of the world's chlorofluorocarbon emissions and 74 per cent of carbon dioxide emissions from fossil fuels.[2]

Yet they made up only 22 per cent of the world's population in the late 1980s. So, in terms of impact per person, the disparities are even more glaring. The typical person in a developed country uses five times more fertilizer, twelve times more oil, and twenty-four times more natural gas than their developing country counterpart. They emit eleven times more carbon dioxide from fossil fuels and twenty-six times more chlorofluorocarbons. They produce forty times more industrial waste, fifty-two times more industrial effluents, and seventy-five times more hazardous waste.[3]

Because of these inequalities continued population growth in the North is at present even more of a threat to the global commons than in the South. The average Northerner emits perhaps twenty times more water and climate pollutants than the average Southerner. If relative consumption and waste output levels stay the same, the 57.5 million extra Northerners expected during the 1990s will pollute the globe more than the extra 915 million Southerners.[4]

But we cannot freeze the picture in the present. The balance is changing.

For most resources the growth rates of Northern consumption per person are slowing. Population growth is also slowing. Environmental awareness and commercial pressures are reducing the environmental damage done for each unit of consumption. In developing countries, meanwhile, populations continue to grow rapidly. Consumption per person, starting from a low level, is also growing. And many Third World countries are industrializing rapidly.

Within just a few decades the developing countries will be the biggest polluters in many areas. In 1977, for example, developing countries were using only 27·5 per cent of world fertilizer production. By 1988 their share had risen to 42 per cent. On these trends, developing countries will be the majority consumers of fertilizer by the mid 1990s. By the end of the century they will account for 60 per cent. Their share of carbon dioxide emissions from fossil fuels is projected to rise from 26 per cent in 1985 to 44 per cent in 2025. Their share of emissions from deforestation is already 100 per cent, since Northern countries are increasing·their net forest area.[5]

However, the old dichotomy between 'developed' and 'developing' countries is increasingly obsolete. Third World countries are found at every stage of income and industrialization: from Guinea, where only 3 per cent of Gross Domestic Product came from manufacturing in 1989, to Brazil, where the share was 35 per cent – higher than Germany or Japan. From Ethiopia, where real income per person, in internationally comparable dollars, was $330 in 1989, to Hong Kong, where it was $15,660 – higher than all but 5 developed countries.[6]

And there is considerable income inequality within countries. The élites of almost all developing countries are consuming and polluting at middle-class Western levels. What differs is only the size of the élite.

The rich use more resources, wherever they live. In Mexico

City, the residents of high-income Chapultepec district use nine times more water per person per day than those of low-income Nezahualcoyotl. The rich discard more wastes. In Port Harcourt, Nigeria, the most profligate district throws away five times more municipal wastes per person than the least.[7]

Rich families contribute far more to global warming than poor. We can estimate roughly how much if we assume that fossil fuel use follows the same pattern as income distribution. Then in the United States, for example, the richest 10 per cent emitted 12·6 tonnes of carbon dixide per person in 1987 – eleven times more than the 1·2 tonnes of the poorest 20 per cent.[8]

This exercise produces some intriguing results. Carbon dioxide emissions in the United Kingdom and Japan were only 0·8 tonnes per person among the poorest 20 per cent. This is on a level with the national average for Chad and Iran, and is less than the average Zimbabwean.

By contrast, the wealthiest 10 per cent of Malaysians emit 2·4 tonnes each – on a par with the European average. The wealthiest tenth of Colombians warm the globe as much as the typical Swiss. The richest one in ten Brazilians emit more CO_2 than the average Frenchman.

In terms of the global commons, the poor in developed countries probably do little more damage than the average person in developing countries. Rich Southerners, meanwhile, do just as much damage as the average Northerner.

Destitution and degradation

But what about the other half of the cliché? Are the poorest people on earth really among the great environmental destroyers? Are they really more likely to degrade their environments than the great mass of people who are neither rich nor very poor?

In some respects they are. They are more likely to gather free fuels. In deforested areas it is the poorest who exert most pressure on the remaining trees and shrubs. It is the poorest who gather every last dried piece of dung for fuel, instead of leaving it to fertilize the soil. Among farmers, it is the poorest families where males are most likely to migrate for work, leaving the wife at home too burdened to take on any extra work to conserve soil. In all these cases, poverty works through the technology element of the three basic factors.

But bigger farmers also degrade the environment, often in bigger ways and always over bigger areas. They are more likely to use tractors, which can damage sensitive soils. They are likely to own more livestock – and if cattle and goats are not properly managed, they can do more environmental damage than humans.

In Lesotho, the poorest 17 per cent of people possess neither fields nor livestock. Since they have no access to land, they cannot degrade it. The next poorest are the 28 per cent who possess fields but no livestock. They degrade their land by not fertilizing or conserving it. But so do the better-off, and they have more land per person. But those who do most damage in Lesotho are the 47 per cent of households who own cattle and, among these, the wealthiest 23 per cent of households who own 74 per cent of the cattle. Livestock degrade the highlands in the summer months. In winter they eat stubble and trample down terrace edges.[9]

In general, there is little reason to believe that poor smallholders are less likely to conserve their land than owners of larger plots. The reverse may be the case. Yields are usually higher on small farms than on large. Vegetation cover will be thicker, giving better protection against wind and rain. And smallholders are more likely to conserve every inch of land when it is their only resource.

The poor are no more likely than the somewhat better-off to start farming in rainforests or marginal areas. Surveys of migrants to cities and to other rural areas show them to be younger and better educated than average. So they are likely to come from families that are not among the poorest. The tendency to move into forest or marginal land is more a generational matter than a rich–poor one. The most likely people to clear new forest, or try their hand in a marginal area, are newly married couples, whose own parents have some land but not enough for an extra household. In humid African countries the better-off are most likely to get hold of large concessions of forest land, to clear and farm with hired labour or tractors.[10]

The poorest families of all, in most parts of the world, are those of widows, divorcees and single mothers. These are least likely of all to open up new land.

There is some truth in the association of poverty with environmental degradation inside marginal areas (see pp. 126–132). But causation is not one way here. Even before it is degraded, a marginal area by nature does not usually produce enough surplus to lift its inhabitants out of poverty. Poor areas and poor people destroy each other.

In cities, the poor live in areas with the worst environmental hazards (see Chapter 12). But they do not cause the problems: they suffer them. The shanty town as such is not an environmental problem. The problem is air pollution and lack of clean water and sanitation. These are just as often the result of bias in city spending as of poverty.

It is often suggested that the poor have more children. If so they would degrade the environment through the population element of our three leading factors. Many aspects of poverty do push towards higher fertility: high death rates among children; lack of education; inability to afford family planning; absence of social security in old age. But within any particular social group, the poorest tend to have fewer children than the somewhat better-off. In landowning families, the number of children tends to rise as farms get bigger, because the farmer needs more labour to work the land, and can produce more food to rear more children.[11]

Finally, consumption and waste per person is also lowest among the poorest. All in all, the poor probably tread lightest of all upon the earth, and do less damage to the environment than any other group. They are victims, not perpetrators.

Dimensions of inequality

Inequality in the distribution of land has a strong impact on the environment. As we saw in Chapter 9, there are many parts of the world where expropriation has driven large numbers of people into marginal or smaller areas. But this factor works its impact on the environment through population, by artificially increasing population density in certain areas.

International inequalities work mainly through technology.

Low and unstable commodity prices lower the value of land, making conservation less worth while. High levels of debt in the 1980s forced governments in Africa and Latin America to cut imports of fertilizers and reduce their agricultural extension services. Debt slowed the spread of new technologies to increase yields and improve conservation.

Inequality between the sexes also affects technology. In Africa women do an estimated 70 per cent of agricultural work, on top of fuel- and water-gathering and grinding. They may be short of time for additional work involved in conservation. Even so, women heads of households are the backbone of Kenya's vigorous soil conservation programme.

Sexual inequality promotes higher population growth rates. Birth rates are highest in the regions where women's status is poorest – in Africa, Moslem countries and the northern part of the Indian subcontinent. Here women usually depend on sons for security in divorce or widowhood. Hence there is an incentive to have more sons, and to keep trying for boys if daughters keep arriving. In these regions female enrolment in school is much lower than male.[12]

Imperfect markets

The radical right raises a contrasting set of issues relating to freedom of markets and private ownership of property.

Free-market price signals have worked reasonably well in regulating our use of mineral resources. If a mineral grows scarce, its cost of production rises. This stimulates recycling, reduced consumption, use of subsitutes and cheaper production methods. The free-market mechanism ensures that, as the resource depletes, it is more carefully husbanded.

Improper interference in free-market prices often results in environmental damage. State control of crop prices in Africa, where they have been artificially low, led to underproduction. It also led to undervaluation of land, so conservation was less economical. In Europe, crop prices are fixed above world market prices. The result is overproduction, wasteful use of chemical inputs, fertilizer and pesticide pollution and loss of bio-diversity.[13]

Command economies, with fixed prices, no competition and no profit motive, use resources more inefficiently. In 1985, to produce each $1000 of GNP, China used more than twice as much energy as North America. The USSR and Poland used three times more than Western Europe.[14]

Yet the perfectly free market has imperfections of its own.

Free-market prices reflect supply and demand, costs and benefits between parties to the transaction. They don't reflect costs and benefits to third parties, other species or future generations. They don't account for the costs of pollution or resource depletion. Hence they don't help to regulate our relationship to the environment as sink for wastes.

So government interference in market price is justified to ensure that the full environmental costs are reflected in the price of a product. This includes the costs of reducing pollution to acceptable standards, costs borne by sufferers of pollution or depletion, benefits forgone by future users, and so on. When this is done, the market mechanism will begin to help regulate our use of the environment as sink for wastes.[15]

The supposed tragedy of the commons

The way in which natural resources are owned and controlled has a strong effect on the level of environmental damage.

Land is best cared for when it is in the freehold of the person who operates it. Owner-operators have an incentive to make sure that their land continues to produce for their lifetime – and that of their children. Small-scale private ownership encourages sustainability.

Any weakening of the link between the operator and the land leads to a lower level of care and conservation. That includes not only state ownership and collective farming, but absentee landlordism, sharecropping and insecure tenancy. Wherever farmers cannot be sure that they or their children will reap the benefits, they will not plant trees or conserve the soil.

Many kinds of resources are not privately owned, but open to all comers or all members of a community. US environmentalist Garret Hardin has argued that common ownership of commons is the cause of much environmental degradation.

Envisage a pastoralist grazing livestock on a common rangeland. He wonders whether to add an extra animal to his herd. If he does so, the range will degrade a little faster than otherwise. But the individual herdsman will suffer only a small proportion of this loss, since it will be spread across the whole range. Against this he will reap 100 per cent of the benefit. The only rational decision, then, is to add the extra animal. And another. And another. And every other herdsman reaches the same decision.

This is Hardin's well-known tragedy of the commons: 'Each man is locked into a system that compels him to increase his herd without limit – in a world that is limited. Ruin is the destination toward which all men rush, each pursuing his own best interest in a society that believes in the freedom of the commons. Freedom in a commons brings ruin to all.'[16]

Hardin applied the theory to all commons used as resources. In the case of forests or fisheries, the decision is whether to cut an extra tree or catch an extra fish, weighed against the risk of depleting the resource base. In such a case the first man to a tree with an axe, the first to the lake with a net, will reap the full benefit immediately. Everyone else will lose.

The theory was also extended to waste sinks – what Hardin called 'the tragedy of the commons as a cesspool'. It costs someone more to purify waste before releasing it than they stand to lose from the additional pollution. The 'rational' decision is to pollute. Here too, says Hardin, we are locked into a system of fouling our own nests.

It is a bleak and cynical vision: a war of all against all, where each person follows their self-interest down the road to social chaos. If it were inescapably valid, the human race would have died out long ago.

The real tragedy of the commons

In practice, things don't usually work out this way. As population density increases, the way in which natural resources are owned or controlled changes too. Completely open access gives way to common property. Resources are owned collectively by the tribe or village, and regulated to avoid abuse,

overuse, or encroachment by outsiders. At even higher
population density, private property develops in farm land.

Pasture and forests remain as commons for longer. In some
cases they too are privatized. In others they remain community
property, or are nationalized as state property. Fluid resources
such as big rivers, oceans and atmosphere cannot, by nature, be
privatized at all. Other resources – particularly semi-arid
pastures – should not be privatized. Their condition fluctuates
from year to year, and the survival of herds and herders
depends on freedom to move into other areas.

Even where common ownership persists, the tragedy of the
commons need not apply. For, as the critics of Hardin's view
have pointed out, people co-operate to solve the problems
arising on a commons. 'Rational men do not pursue collective
doom,' British livestock expert Stephen Sandford remarks.
'They organize to avoid it.' Hardin forestalled some of the
criticism, pointing out that people can exercise 'mutual
coercion, mutually agreed upon'. They agree to laws against
murder and jumping red lights. They can also agree on
institutions and regulations to preserve commons.[17]

And they do so agree. Most pre-modern societies have
arrangements to govern sharing of common resources, and to
prevent their degradation. African pastoralists have elaborate
kinship arrangements for spreading herds widely across the
ranges. In Bali canal water is not simply snatched by the farmer
nearest to the headwater. It is distributed equitably by the *subak*
or irrigation committee. For every unit of land he owns, a man
has the right to a certain measure of water flow. Anyone who
takes more is guilty of water-stealing and can be heavily fined.
Everyone has to contribute to the upkeep of the canals.[18]

The real tragedy of the commons, then, is not the existence of
open access or common property in resources. These forms of
control are perfectly appropriate in situations of low population
density. And for resources that by nature cannot or should not
be privatized they are unavoidable.

The tragedy of the commons arises when institutions of
ownership or control fail to move in step with the prevailing
level of population density. In most of Africa, population
density has already reached the level at which private
ownership of land would be a normal development. But in most

countries land is nationalized. The state has blocked the transition to private ownership of arable land.

In many developing countries, forests, too, have become state property. Local communities have lost power to control the forest, or to benefit from its conservation. Yet corruption, lack of resources or personnel, mean that the state cannot take care of it sustainably. And so the forest degrades.

In other cases communal institutions once existed, but were destroyed by colonial powers or modern states. Many Indian villages maintained their commons by unpaid communal labour, and taxed the users of commons. But local taxes and labour obligations were suppressed by the government. In Africa pastoral groups controlled access to rangeland by guarding the wells they owned. When governments sank public wells, everyone had free access. The rangeland was opened to all comers, and desertification began around the waterholes.[19]

However, some degree of scarcity or degradation must arise before people see the need to control use of the resource. Further depletion will occur in the time taken to develop appropriate institutions. The bigger the commons, the longer this will take. And during the delay the commons degrade.

The power of democracy

Sooner or later, people perceive environmental degradation as a problem and act on it. The amount of damage done depends on the length of the delay (pp. 249–54).

The fastest feedback occurs where the person who suffers from degradation is in a position to control it directly. This is why private ownership of land produces a faster response than common ownership; and why local community control works quicker than distant bureaucracies.

In most parts of the world women are first to notice environmental degradation, first to be badly affected. Women clean the dirt of air pollution and nurse the sick children. As forests disappear and wells dry up, women have to walk further and further to get wood and water. Yet they usually lack the power to act on their problems. Men control farmland and common land. Men decide if trees are to be planted or not.

Where women are in a position to act – or where men share the
burdens that make women feel the problems earlier –
environmental degradation can be dealt with much sooner. If
African men had to gather fuelwood, or if African women were
free to plant trees, deforestation and fuelwood shortage in Africa
would be remedied much sooner.

Many environmental problems are the result of what
economists call 'externalities': the effects on third parties with
no direct means of controlling the activities that cause the
problems. Democracy and the rule of law are the only way in
which externalities can be brought into consideration. People
must be able to protest, demonstrate and organize when their
interests or health are damaged. That means guaranteeing
rights of free speech, assembly and association. The media must
be free to report protests – and to investigate pollution
incidents, so they must enjoy complete editorial freedom from
government and commercial pressures. There must be free
access to official information – including the results of official
inspections of manufacturing processes. There must be free
elections with multiple candidates, so that representatives who
don't respond to public pressures can be booted out. There
must be an independent legal system with equal access for the
poor. Many 'democracies' do not meet these requirements.

It is no accident that Communist countries were among the
most polluted in the world. Where farms and factories are state-
owned, criticism or protest is suppressed as dissent or rebellion.
Independent organizations are seen as a potential focus of
resistance, and suppressed.

Even where full democratic rights exist on paper, they may
not be effective. Where there is widespread illiteracy, people
don't know how to assert their legal rights, to collect evidence,
to publicize their case. Ballot-rigging is common. And wherever
there is corruption, nepotism or cronyism, protests about
pollution will be ignored.

The story of Cubatão, near São Paulo's port of Santos in
Brazil, illustrates just how important democracy can be. Once a
pleasant town by a mangrove-lined valley, cradled among tree-
covered mountains, Cubatão developed in the 1960s and 70s
into a centre of heavy industry. Steel, fertilizer and chemical
factories were built. Poisonous wastes were dumped in the

rivers and pumped into the air. Workers built shacks on hillsides and swamplands.[20]

By 1985 the mangroves were gone, and the hillside trees were skeletal. Erosion from the unprotected slopes silted up the rivers. Fish died out. There were high levels of tuberculosis, pneumonia, bronchitis, emphysema, stillborn and deformed babies. Cubatão became known as the Valley of Death. Yet under the military regime the press was banned from reporting conditions there. Right-wing dictatorships are no better for the environment than left-wing ones.

Everything changed when democracy was restored. In his election campaign, state governor Franco Montoro pledged to reduce pollution. From 1983 on a massive clean-up costing over $300 million was launched. By late 1988 249 out of 320 sources of pollution in Cubatão had been controlled. Smoke had been cut by 92 per cent, sulphur dioxide by 84 per cent, industrial effluents by 91 per cent. Fish returned to the river after a thirty-year absence. The hillsides were reseeded with trees.

Towards a holistic analysis

All these direct and indirect causes of environment problems work together. In the real world they weaken or reinforce each other, interact and influence each other.

In communist and other centrally planned countries, environmental degradation has been particularly bad because of a combination of factors: no market freedom, no competition, no private property in resources, no democratic rights. Land degradation is worst in Africa because rapid population growth has coincided with technological stagnation, under one-party and military dictatorships, with state ownership of land and forests and government control of prices. Deforestation has been so dramatic in the Amazon because of unequal landownership, the absence of democracy for much of the time, resistance to land reform demands, and fiscal favours to rich ranchers.

Our village studies have shown just how complex are even the major causes in individual cases. In Madagascar, deforestation was due to population growth coupled with technical stagnation, which in turn was reinforced by poverty and

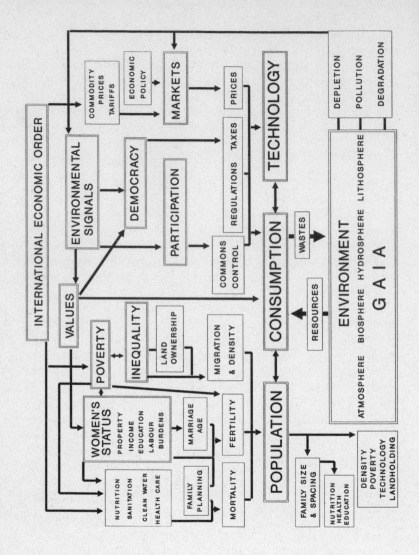

international debt. In Burkina Faso desertification was the result of population growth and climate change, plus a series of obstacles that together amounted to a technological trap. In Abidjan rapid urban population growth – speeded along by rapid rural population growth – combined with too costly technology and biased city spending to keep many families in squalor. On Hatia Island exposure to disaster was due to population growth on limited land resources, unequal land distribution and river erosion. (See Chapters 5, 10, 11 and 16.)

One-sided analysis leads to one-sided solutions. Academics and planners must learn to look at all aspects of the situation, and to include population every time. The challenge ahead is so great that we can save the earth's inheritance for future generations only if we understand and deal with all the interlinking causes.

Chart One (opposite): Major population and environment links. The human-Gaia interaction is a system. All factors interact: there are no 'ultimate' causes, though some (double-outlined here) reach deeper and wider than others.
Bottom: Population, consumption and technology operate *directly* on the environment, extracting resources and emitting wastes. When resource demand or waste density exceed carrying capacity, depletion or degradation follow.
Right: This creates problems – environmental signals. When these are perceived, they alter prices, or, through politics, lead to new taxes, regulations etc. At crisis points they also affect values. Consumption and technology change in response. Market prices can be distorted by local or international economic systems and policies. The absence of democracy, local participation, or effective control over commons, also delays response.
Centre: Inequality, especially of land ownership, deepens poverty and speeds migration, increasing population density in parts. Values, deriving from religion and culture, influence women's status, consumption and political factors.
Left: Poverty tends to keep fertility high. It reduces the ability of families and government to provide good health and nutrition, and hence raises mortality. Women's education, income and burdens affect family welfare, age of marriage and use of family planning.
Some population cross-links are shown bottom left. Many other factors and cross-links are not shown for clarity reasons. See Chapters 17 and 18 for full analysis.

19

WE DEFY AUGURY:
options for action

On the eve of the Third Millennium, we are in the embrace of an environmental crisis that is coiling around more and more regions and ecosystems. Accelerating deforestation in the South, forest death in the North, red tides, the ozone hole, the threat of global warming: all have arrived over the space of a mere fifteen years. Underlying them is the long attrition of biological diversity, and the progressive degradation of land.

These problems arose when perhaps no more than 1·4 billion people were consuming at levels of at least moderate material affluence.[1]

Yet ahead lie four decades of the fastest growth in human numbers in all history.

Another earth

Over the next decade as many as 97 million people will be added each year, according to the United Nations medium projection. But this headlong growth will not stop at the end of the century. In the following twenty-five years annual additions will still be running at an average of 90 million a year – every decade an extra India. And for the quarter century after that there could well be another 61 million – an extra Britain, France or Italy – every year.[2]

After 2050 the slowdown will be marked. But the final totals will be massive. The United Nations projects that world population will reach a plateau, some 150 years from now, of around 11·5 billion. This is a whole extra human world on top of 1990's 5·3 billion, plus an extra China thrown in for good measure.[3]

If these projections prove correct, the world of 2150 will look very different from the world we have been used to. Although the developed countries behave as if they make up more than half the global population, they were only a third back in 1950. Today they account for just under a quarter.

Over the next century and a half, more than 98 per cent of population growth will be concentrated in the developing countries. For every extra person in the Northern countries, there will be sixty in the South. In the world of the twenty-second century Southerners will outnumber Northerners by nine to one.[4]

The geopolitical shifts will be momentous. The dominance of the West and North will be history. By AD 2050 India will have overtaken China as the world's most populous country. Eventually there will be 1950 million Indians to 1390 million Chinese. Populations in Latin America and Asia (excluding China) will rise by two and a half times. The most explosive growth of all will be in Africa, whose population is projected to grow from around 642 million in 1990 to 3090 million in AD 2150. By then Africans – in 1950 only two-thirds as numerous as Europeans – will outnumber their Northern neighbours by seven to one.

I use the future tense for simplicity. But these projections are not predictions. They assume, for example, that by the 2020s the average numbers of children per mother in developing countries will have dropped to the level of developed countries in the 1960s. This will only happen if there are steady improvements in health, education and the availability of family planning. And these improvements will have to be fought for.

In the 1980s they were not fought for. In 1984 the United Nations expected the typical number of children per mother to drop by two in the fifteen years from 1970–75. In the event, they dropped by only 1·5. And so the projections made at the beginning of the decade looked rosy towards the end. The medium projection for the year 2025 had to be raised, from 8·2 billion to 8·5 billion. An extra 300 million people, another Soviet Union, thanks to one decade's procrastination. The UN long-term projection made in 1991 was 1·4 billions higher than the one made in 1980 – an extra North and South America plus Europe.[5]

If similar slippages recur, world population could all too easily rise to a peak of 13 billion or more.

The consumption explosion

Our present environmental problems arrived, too, after only four decades of mass affluence in the Northern countries.

It is easy to forget just how briefly the consumer society has existed so far. Though it can be dated, in America, from as early as the second decade of this century, recession and war delayed its arrival elsewhere until after 1945.

Take a typical developed country, Britain. In 1931 less than a third of British homes had electricity. By the late 1940s, the share had risen to 86 per cent. But even then there were only two functions that all electrified homes enjoyed – light, and ironing. Only one in twenty-five households owned a washing machine. One in fifty had a refrigerator. In 1950 there were only sixty-six motor vehicles per 1000 people – one-seventh of the 1988 ownership level.[6]

But by 1960 the consumer revolution had spread to encompass most of the Western world. Now it is reaching an expanding middle class in developing countries. The consumer revolution continues – and the pace is accelerating. We face a consumption explosion at least as momentous as the population boom.

Bikes and battery radios come first. Electrification opens the door to an unending succession of appliances. The real price of these, relative to incomes, has been falling. Ownership in developing countries reaches higher levels, at lower incomes, than in Western countries in the past.[7]

Refrigerators are an early purchase. In the mid 1980s, 13 per cent of even the poorest households in Kuala Lumpur owned a fridge. In Manila 15 per cent of low-income families owned one. In Kuala Lumpur's second lowest income band, one family in two owned a fridge. Ownership in both cities ranged right up to 96 per cent among high-income groups. In the Chinese capital, Beijing, almost two-thirds of all families owned a fridge.[8]

Televisions are next in the scale of priorities. I have seen shanties in Lantao island, Hong Kong, with television sets in the corner amid bare floorboards and spartan furniture.

Television is consumerism's Trojan horse. It multiplies material wants directly, through advertising, and indirectly, through soap-opera life-styles. Many Third World TV programmes are imports from America, the richest consumer society in the world.

Cars have the biggest environmental impact of any material possession. The world vehicle population has grown more than twice as fast as the human. Between 1950 and 1988 the number of cars, trucks and buses grew at 4·1 per cent a year, while human numbers grew at only 1·9 per cent.[9]

Developing countries owned only 79·5 million vehicles in 1988 – just under 15 per cent of the world fleet of 540 million. They had only 21 vehicles per 1000 people – against 385 per 1000 in developed countries. But their share is growing.

In rapidly industrializing countries the speed of expansion is terrifying. The total number of vehicles in South Korea grew from 129,000 in 1970 to 2·04 million in 1988. In 1988 alone an extra 424,000 vehicles hit the road. The annual growth rate in vehicle numbers averaged 16·6 per cent over that period – more than ten times faster than population. Population growth alone accounted for only 7 per cent of the additions during the 1980s: increased per capita ownership for the remaining 93 per cent.

India's fleet doubled in size to 3·3 million in the eight years from 1980, an average growth rate of 9 per cent a year. New Delhi, staid and quiet only a decade ago, now buzzes with swarms of motor scooters. China has been adding over 600,000 vehicles a year in recent years. And both countries, with only four vehicles per 1000 people in 1988, are only just approaching the foothills of a steep climb. If the Chinese owned vehicles at European levels, their vehicle fleet alone would equal 70 per cent of the present world total.

The combined impact of population and consumption growth is explosive. In a world of 11·5 billion people, with car ownership at 1988 developed country levels, there would be 4·5 billion cars, trucks and buses – more than eight times the present number.

Abolishing underconsumption

What can be done about the double explosion of population and consumption?

Environmental impact is the direct outcome of population, consumption and technology working together. So there are three possible avenues for reducing environmental damage. We can slow population growth, with a view to an eventual reduction over the very long term. We can reduce consumption per person. Or we can adapt technology, to meet our increased needs with lower environmental impact. And we can combine all three.

Let us examine the prospects for each one of these routes separately.

Consumption is the hardest to tackle. For here we are dealing not just with the overconsuming rich. There are an estimated 1·1 billion people in the world who are under-consuming: people in absolute poverty; people who don't earn or produce enough to meet their basic needs in food, clothing, shelter, fuel, and so on. Many of their social needs are unmet too. They are victims of multiple deprivation. They include the 1 billion adults who were illiterate in 1990. Their children are among the 300 million who do not attend school. They are there among the 1·5 billion with no access to health services or safe water, and the 2·2 billion without adequate sanitation. The poorest of them make up the 500 million-plus malnourished.[10]

Any conception of a sustainable world must include an end to absolute poverty and all its component aspects.

This means increasing the consumption of the absolutely poor. But this could be achieved with no great impact on the environment. A sketch must suffice, and it must inevitably oversimplify.[1*]

Much malnutrition is the result of illness, which reduces food intake and absorption. Improved water supply and sanitation can help here. Apart from that, malnutrition primarily affects people who don't have enough land to grow the food they need, or enough income to buy it. Ending malnutrition therefore involves a double strategy of providing land and farm inputs, and creating jobs. In Latin America, the Near East and some parts of Asia, much land can be provided by redistributing large holdings. Elsewhere intensification efforts focused on

[1*] The reader is referred to my other books, *The Third World Tomorrow* (Penguin, 1980), and *The Greening of Africa* (Penguin and Paladin, 1987).

smallholders can increase production. For those who remain landless, extra jobs can be provided by encouraging small-scale labour-intensive enterprise, rather than large-scale capital intensive manufacturing, and in rural areas and small towns as well as big cities.

The social aspects of absolute poverty – lack of health services, water, sanitation and education – can be alleviated at even lower environmental cost, by reorientating spending priorities and using low-cost appropriate technologies.[11]

Reducing overconsumption

These low impact strategies rely on redistribution of assets or services. They relieve the rich of some of their excess, in order to provide for the poor. It is easy to call for, but not at all easy to achieve. The rich do not, as a rule, stand by and let themselves be robbed. They resist, often violently. They pull every political lever within their grasp.

Poverty has more usually been alleviated by overall economic growth, rather than by redistribution. This route takes longer to abolish poverty. And it involves increasing everyone's consumption, not just that of the poorest. Inevitably, the environmental cost is higher.[12]

But no one in the Third World would agree that development should just eradicate absolute poverty and then stop. When Third World people talk about developing, they are talking, explicitly or implicitly, about joining the feast. They want their turn at the top table. Nothing can, will, or should stop them.

The call to reduce overconsumption is also easy to make. But we must first define it. Yesterday's luxuries are today's social expectations. The basic items of the Western inventory are not artifical 'wants' rather than needs. Electric light extends active hours and increases earning power, by opening up the night. Refrigerators prevent food wastage and contamination. Indoor taps, showers and toilets improve health as well as convenience and privacy. Room heating is essential for health and survival in cold climes. Air conditioning raises work productivity in hot countries.

As affluence increases, more and more inessentials are added to the list. To move with less effort and more freedom, we buy cars. For information and stimulation we acquire TVs, video recorders, record, tape and compact disc decks. For comfort we buy soft furnishings. To save effort we buy machines that reduce domestic labour: washing machines, dishwashers, tumble dryers, vacuum cleaners, cookers, grinders, mixers, lawn-mowers, hedge trimmers. The time these appliances save is spent using more commercial energy, on electronic games, sports, holidays to increasingly distant destinations. The whole thrust of development is to replace human labour with other sources of energy, and to increase mobility and variety of direct or vicarious experience.

Where do the limits of untrammelled consumption lie? In 1990, for every 1000 Americans, there were 811 television sets – one for each person over twelve. There were 631 cars per 1000 people – one for every adult over twenty-four. Nine out of ten US households own at least one car. Some 37 per cent own two cars, while 18 per cent own three or more.[13]

And this is just the stock at one time. The inventory is continually upgraded. Out of 190 million registered motor vehicles in the USA in 1990, 23·4 million were new that year. The average age of the whole fleet was only eight years. Fashion and innovation encourage continuous replacement of goods. People throw out clothes, kitchens, furniture, cameras, computers, long before they are worn out or bust, because more fashionable or more efficient products have appeared.[14]

Conspicuous consumption is another stimulus to overconsumption. Ever since humans have been capable of producing a storable surplus over and above their immediate needs, they have sought to impress each other with the sheer quantity of their material possessions.

Kwakiutl grandees on Vancouver island used the potlatch. They built a great bonfire of blankets, canoes and other treasured possessions, then burned them in front of their rivals, to show they could spare them. To save themselves from shame, the invited guests had to lie unmoved in their places, though the fire blistered their shins or blazed up and set the rafters alight. The host, too, had to show the most complete indifference to the threatened destruction of his house.[15]

The modern equivalent are personal jets, luxury yachts, multiple mansions. But today's grandees keep their possessions, and burn up the earth to accumulate more. There seems to be no upper limit to the level of individual consumption.

Dematerialization

Some observers have found encouragement in recent trends in the rich countries. Less energy and less material input – measured by weight – is being used to create each unit of national wealth. Economists talk of the 'dematerialization' of Western economies, of a new long wave of growth in the world economy. In future people will spend a greater share of their incomes on services rather than manufactures; on quality rather than quality; on 'positional' goods like well-sited housing rather than sheer quantity. Products will contain a higher content, by value, of information and software and relatively less material. Third World countries, rather than following the outdated path of the West earlier this century, will adapt to the new wave.[16]

Dematerialization is something of a mirage. The proportion of material goods to total spending may well be falling in many Western countries. But incomes are rising, and Westerners own more material possessions of more kinds than ever before. Though their composition and weight may be changing, it seems unlikely that the total volume of material resources used is falling.

Much has been made of the increase in energy efficiency from about 1973 onwards. This was precipitated by the rise in oil prices. Energy use per unit of national income fell in 54 countries out of 147 between 1977 and 1987. But meanwhile incomes per person were growing in most countries. So only 41 countries managed to reduce energy use per person over this period. And of course the number of persons was growing almost everywhere. The end result was that only 22 countries cut their total energy use. Most of those were debt-hit countries in Africa and Latin America.[17]

Over this period hailed for its growing 'energy efficiency' 125 countries increased their overall energy consumption, and world energy use expanded by 20 per cent.

And when oil prices began to fall significantly after 1986, some of the gains began to evaporate. World energy consumption, which had grown at only 2·4 per cent in 1986, rose by 3·1 per cent the following year and a massive 3·7 per cent in 1988.[18]

278 THE THIRD REVOLUTION

Another symptom of alleged dematerialization has been the drop in steel use in developed countries. This has been steeper and more lasting. There is a global trend towards lighter and smaller products. Plastics, ceramics, lightweight alloys, glass fibres, are replacing heavy metals in many uses. But the trend has not been enough to halt the rise in global steel consumption, which was 5 per cent higher in 1988 than in 1973. Much of the dematerialization of developed countries is due to the shift of basic manufacturing capacity to the Third World. While America's steel consumption per person dropped 39 per cent between 1973 and 1988, China's rose by 141 per cent. [19]

The prospects for reducing consumption are not bright. As long as some people overconsume, others will aspire to emulate them. The Romans limited expenditure on banquets, banned the use of golden tableware, prohibited the wearing of silk by men. But it seems unlikely that democratic governments could bring in such sumptuary laws to limit consumption, or reduce inequality to the point where no one could afford gross overconsumption.[20]

Only two developments could alter the outlook. One is the spread of new social and environmental values, which could induce a voluntary reduction in consumption (see p. 301ff). The alternative – if this does not work – is a continued build-up of environmental problems, which could eventually make compulsory reductions unavoidable.

Technologies for a sustainable world

In the shorter term, it will be easier to reduce the environmental cost of consumption, than to lower consumption per person. This means action on the other two elements of our equation: population, and technology.

Technological change can achieve a huge impact over a relatively short period. Sulphur dioxide emissions in Western nations fell by 38 per cent in just eighteen years between 1970 and 1988, thanks to controls on emissions and shifts to low-sulphur energy sources. Without population growth, emissions would have fallen by 47 per cent.[21]

In China technology changes – mainly increased fertilizer application and improved varieties – raised rice yields by 57 per cent in only fourteen years from 1974. As a result 40 per cent more production was obtained from 11 per cent less area. Despite increased population, less land was needed for farming, and forest land expanded by 2 million hectares.[22]

Technological change could meet the energy needs of development at much lower cost. According to Brazilian economist José Goldemberg and colleagues, the most advanced technologies known today could cut energy use per person in developed countries by one-half, even if standards of living improved by 50 to 100 per cent. For a typical warm-climate developing country, a Western life-style could be provided with about 1 kilowatt per person, only a slight increase on present levels of 0·9 kw.[23]

Many of the technologies needed to conserve forests and soils, and reduce pollution, are already known. To reverse forest degradation by logging we need sustainable management of forests (see p. 92f), and multiple use of the widest range of forest products. To halt deforestation we need agricultural intensification on existing farmland – higher food production from the same area. To conserve soils, farming must move into agroforestry (the inclusion of trees of all kinds on the farmland), into organic inputs of manure and compost, and biological methods of pest control. Human sewage must be returned to the soil, to halt the haemorrhage of nutrients into the sea.

On the urban front, low-cost technologies and support for self-help can provide a far better urban environment, for more people, at lower cost. Recycling solid waste reduces use of resources without reducing consumption. Organic farming will cut the flow of sewage, fertilizer and pesticide wastes into water courses. Reduction of other pollutants can be more expensive. But in the case of carbon dioxide, increasing energy efficiency saves money as well as reducing emissions.

The key problem with low-impact technologies is adoption on a wide scale. They must be socially acceptable and economically attractive to potential users. Many green technologies have fallen down on this last criterion. In Africa, conservation techniques are popular where they are very low cost, and produce high returns on investment. Conservation techniques which cost too much in labour and lost land, such as terracing, are less successful. Cheap techniques which yielded a healthy increase in production – like the stone lines in Burkina Faso (see pp. 149–51) – spread rapidly.[24]

These same criteria can be applied more generally: the lower the costs, the higher the benefits, the more likely it is that an environmentally benign technology will spread. Technologies which increase production or incomes at the same time as they conserve the environment will have the most rapid impact.

Running up the down escalator

Technology change is indispensable. But it is no panacea.

The task ahead is enormous. Consider the case of the galloping car. We look to improvements in fuel efficiency to allow us to carry on driving, while doing less injury to the atmosphere.

Some progress has been made. In Western countries, petrol consumption per car fell by 26 per cent in the fifteen years from 1973. Consumption per kilometre driven dropped even further, by 29 per cent. But these gains were more than wiped out by a rise in car numbers of 58 per cent. The overall result was a rise in total petrol consumption of 17 per cent between 1973 and 1988. And this was during a period with two massive oil price rises.[25]

The rise in car numbers seems unstoppable, even in countries with quite high levels of ownership. The United Kingdom vehicle population, for example, grew from 17·4 million in 1980 to 24·6 million in 1988.[26]

As oil prices began to fall again after 1987, efficiency gains tailed off. In most developed countries there was a shift to more powerful engines, more added extras, more automatic transmissions, more heavy safety equipment, and catalytic convertors – all of which add to fuel consumption.

Shifting to more sustainable forms of energy will be the focal task. But even renewable energies have impacts. There are virtually no ways of meeting large consumption needs of huge numbers of people without signficant environmental losses.

Dams provide energy without atmospheric pollution – yet

they arouse violent hostility from environmentalists because they flood valleys, displace people, destroy wildlife habitats. In 1989 the world was already using over 40 per cent of the estimated hydropower potential. In fact, with dams silting up at an average rate of about 1 per cent a year, dams can be considered as a short-term, unrepeatable stopgap that will within a century or so convert all suitable valleys into alluvial farmland.[27]

How about growing plants for conversion to ethanol? This would require a land area of 600 million hectares even to replace current world fossil fuel use. This amounts to 41 per cent of all the cropland in the world in 1988.[28]

Wind power is promising. But it involves placing windmills 50–100 metres tall – bigger than 400 kilovolt transmission pylons – on exposed hilltops and shorelines where the visual impact will be considerable. Environmentalists are leading the fight against windmills in Britain. We have to trade climate change for loss of unspoiled landscapes: we lose something either way.[29]

Probably the most promising candidate to replace fossil fuels in transport is compressed or liquid hydrogen, produced by solar energy in semi-desert areas from piped-in water. Hydrogen burns cleanly, giving off water again. To replace world fossil fuel use with solar hydrogen would require 53 million hectares of solar panels. But in a world of 11·5 billion people, to supply the equivalent of USA fossil fuel use per person to everyone would take up a massive 615 million hectares – about one-fifth of the world's deserts. And there are many technological problems to be solved in storage, transport and fuel tanks.[30]

This project should have a back, or second

What all this means is that technology change cannot do the job alone. For maximum long-term impact, it must be combined with efforts to slow down population growth.

Population efforts work on a slower time-scale than technology. Tomorrow's parents have already been born: they will grow up and have children whatever we do. And there are more of them than of their parents' generation.

A ball that has been kicked once will continue rolling for some time before slowing to halt. This same principle applies to population. If every woman, starting today, had just enough children to replace her and no more, the momentum of past growth would keep population growing for another half century, adding an extra 2 or 3 billion to the total.[31]

At first the difference between slower and faster growth rates are small in terms of numbers. The United Nations has three projections, with the same starting point in 1985. By 1995, the low projection assumes that each fertile woman will be having 9 per cent fewer babies than on the medium projection. Yet world population will be only 1·5 per cent less, or 88 million people.[32]

But the differences are compound: they build up massively as time progresses. By the year 2025 the low projection has no less than 914 million fewer people than the medium – 11 per cent less. And if world population levels off, some time after 2100, the low projections foresee 4·3 to 5·6 billion – 6 or 7 billion people less than the medium.[33]

If one of these low projections could be achieved, the difference in environmental impact would be enormous.

Deforestation is due to extension of farm and non-agricultural land. Let us assume that, in rough terms, a minimum of 0·15 hectares of cropland is needed per person, with another 0·06 hectares for non-agricultural needs, taken mainly out of agricultural land. Agricultural land will have to expand by roughly 0·21 hectares per person in all.

The medium projection will thus require a total of 12.6 to 14.7 million square kilometres more land than the low projections. If this extra land is taken from marginal areas, it will increase land degradation. If from forest, it will speed deforestation. It is equivalent to 31 per cent of the total forest area of the world in 1989, and double the 6.5 million km^2 of protected natural areas in 1990.[34]

Breathing space

There are big differences in pollution too. Assume that by the year 2100 everyone is producing waste at the European level of

1 tonne per year. Then the extra people of the medium projection will turn out 6 or 7 billion more tonnes of waste than the low.[35]

The capacity of the environment to absorb pollution is limited. In the case of many air and water pollutants we may have to set ceilings to the amount that can be emitted. Because waste carrying capacity is not flexible, these ceilings will be fixed and absolute. They will not increase as the number of persons increases. And it is these ceilings, rather than absolute shortages of resources, that will impose ever-tighter limits on our domestic and economic activities. The more people there are, the lower our pollution 'rations' per person will be.

Emissions would be kept below the ceilings either by reducing consumption – or by spending a higher proportion of income on pollution abatement. This too will mean reduced spending on consumption of other items. Each person in a more crowded world of 11·5 billion people would have pollution allowances 51 to 63 per cent lower than in a world of 4·3 to 5·6 billion.

This may sound like Orwellian science fiction. Yet it is already clear that we may face exactly such a rationing scheme for carbon dioxide. According to the International Panel on Climate Change, human output of CO_2 must be cut by at least 60 per cent if atmospheric concentrations are to be stabilized. This would give an annual global ceiling of no more than 2·8 billion tonnes of carbon content each year.[36]

Even with the present world population, living within such limits would mean harsh restrictions. Each one of us would be allowed to emit only 0·53 tonnes of carbon equivalent per year. This is less than one-third of the present world average, and on a level with Burkina Faso, thirteenth poorest country in the world.[37]

By the year 2025 our individual carbon allowances would have dropped, on the medium population projection, to 0·33 tonnes per person per year – about one-third of the present African average. In a world of 11·5 billion people we could emit only 0·24 tonnes each per year. To reach such a target the UK, for example, would have to cut per capita emissions by 91 per cent from the current level. Americans would have to cut emissions by 95 per cent. Emissions may be covered by tradeable permits, issued on the basis of population. In that case rich countries could buy emission rights from poor. But this would mean less income for other things. And after another century of economic growth, most countries will be seeking to buy permits, while few will wish to sell them.[38]

But if we could achieve one of the low population projections, our carbon allowances would be 0·5 to 0·65 tonnes per person – more than double the allowance if the medium projection comes true. This means more than twice as much freedom of action in all those activities that involve emitting carbon.

Restrictions may be imposed for the other greenhouse gases. Chlorofluorocarbons are being phased out. But their substitutes may be rationed, as well as low-level ozone and nitrous oxide produced by motor vehicles. It will be difficult to impose limits on methane, because of its central role in agricultural production.

Similar limits are already coming into play for fertilizer use in some countries, and this policy will certainly spread. Eventually it is conceivable that a very wide range of liquid and gas pollutants will be covered by such rationing schemes. Water, too, may come to be rationed in many countries, either explicitly or through high prices.

The smaller the world and local populations, the more breathing space we will have. The bigger the population, the more constrained our own lives, and those of all other species on earth, will be.

Aiming low

Slower population growth can therefore make a very significant contribution to slowing deforestation, land degradation and loss of species – and to reaching a stable state with a lower level of irreversible damage. And it could provide the eventual world population with less restrictive limits on emissions of all kinds.

But can we pull it off?

The medium projection assumes that forty years from now the average fertile woman in a developing country will have 2·3 children – down from 4·2 in 1980–5.[39]

The low projections assume much faster falls, dropping to

only 1·83 children per woman in 2020–5, just above the average for northern Europe today. The pace is ambitious, averaging a 19 per cent fall each decade. Yet it is not outside the bounds of possibility. Malaysia, Indonesia, Sri Lanka and Mexico have all achieved drops of more than 20 per cent a decade over two decades. Over the same period South Korea, China, Thailand and Jamaica have seen fertility drops of over 30 per cent per decade. The fertility declines that the low projections require of each region have actually been achieved by at least one country in each region. The only exception is Africa, where the drops of 29 per cent a decade between 2000–5 and 2020–5 are unprecedented – though they have been achieved elsewhere.[40]

The UN's low population projection, then, is both desirable and feasible. It should be adopted as a target, globally, regionally and above all nationally as an important contribution to sustainable development.

But it will demand a concerted effort. How can it be achieved?

First of all, consider how it can *not* be achieved. Compulsion has no role to play. Compulsion over family size is a violation of human rights. Even if it seems justifiable in extreme cases, it is possible only in non-democratic societies. Sooner or later – and these days sooner – democracy is restored, and the inevitable reaction occurs.

Compulsion has been attempted on a wide scale only twice: in India, during the emergency rule of 1975–7, and with China's one-child policy of the 1980s. China's policy will collapse as soon as any measure of democracy is introduced. Couples who have been holding back will quickly make up the lost ground. In India recruits for vasectomy were got by threatening to withhold licences for shops and vehicles, credit, food ration cards, land registration, supply of canal water, job applications; in some cases by police round-ups. Sterilizations ran at 1–3 million a year in the early seventies. In 1976–7 they soared to 8·3 million. But 1641 people died as a result of botched operations, and 525 bachelors were sterilized.[41]

It was largely because of this misguided adventure that Indira Gandhi lost the 1977 elections. Family planning was set back by almost a decade. In 1977–8 sterilizations plummeted to 950,000. It was ten years before the number of couples using modern contraception rose again to their 1972–3 peaks. India's birth rate showed essentially no change between 1977 and 1986.[42]

Compulsion is not only unethical: it does not work except in the very short term. Any measure of compulsion or tough talk of 'population control' is absolutely counterproductive and should be avoided at all costs.

It used to be thought that economic development and urbanization were preconditions of slower population growth. This view is now discredited. Success stories in Sri Lanka, Kerala and Indonesia prove that slowing population growth is possible at low levels of income and urbanization. Conversely, many countries with high incomes and urbanization levels have not seen population slow down as expected.

The case of Thailand leaves the economic development theory in ruins. The number of children per fertile woman plummeted from 6·14 in 1965–70 to only 2.3 in 1987. Use of modern contraceptives shot up from only 14 per cent of couples in 1969–70 to 67 per cent in 1987. Yet as late as 1989 Thailand was only 22 per cent urbanized. Two-thirds of the population worked in farming.[43]

What distinguishes Thailand is its success in the human aspects of development, especially for women. In 1989 only 34 children per 1000 died before their fifth birthday – half the average for Thailand's income group. Female illiteracy was only 12 per cent – almost a third of the average. Girl's enrolment in primary school was 90 per cent of male.[44]

Opening up options

Reducing population growth is not just a matter of dishing out contraceptives. It involves a broad spectrum of measures, hinging around children's health and education, and women's freedom to decide their own destinies.

Agrarian societies induce fatalism: the majority are powerless in most areas of their life. Food supply hangs on the weather. Health and survival depend on the invisible forces of disease and accident. Women's horizons are hemmed in by the burden of labour in home and field. In many societies they cannot inherit or control land. Their role, status and security, depend on childbearing. And so they marry young, start childbearing young, and continue until the menopause.

Children are a double insurance: against the parents' destitution in old age, and against the chance of losing the child itself. Women must be sure that one or two sons will survive to take care of them if they are widowed. Where child mortality is high, this means having three sons, and on average at least six children. The typical fertility for traditional societies, around seven children per woman, represents not just indiscriminate breeding, but the result of careful strategy.

Rapid population growth happens, in simple terms, when death rates start to fall faster than birth rates. Most of the decline in death rates is among children. In 1965, one child in every eight in developing countries would die before its first birthday. By 1989, only one in fifteen. But parents do not respond to the raised chances of their child surviving with an equal lowering of family size. At first they over-insure.[45]

As child mortality falls further, the role of fate and chance dwindles. People's sense of control over their own destiny grows. They can plan the number of children they want, with a very good chance that all will survive. They no longer need to over-insure. At this point fertility starts to decline. It can fall rapidly where education and family planning are widely available.[46]

Education expands the choices available to women beyond childbearing, readies them for cultural change, increases their sense of control over their lives. Women with four to six years' education have, on average, 0·8 fewer children than those with no schooling at all. Women with seven or more years' education marry, on average, over five years later and bear 2·3 children less than uneducated women.[47]

Yet female education still has a long way to go in the regions with fastest population growth. In sub-Saharan Africa only three out of five girls were enrolled in primary school in 1988, and in South Asia three out of four. The other regions approached 100 per cent primary enrolment. But female enrolment at secondary level – which is decisive in reducing fertility – ranged from only 14 per cent in Africa and 26 per cent in Asia, to 52 per cent in Latin America.[48]

The right to choose

Another crucial element in slowing population growth is the availability of family planning: the option to choose family size by effective and convenient methods. Women who have easier access to modern contraceptives tend to use them more, and those who do use them have fewer and healthier children. Government family-planning programmes have a strong impact on birth rates. In countries with upper-middle levels of development and a strong family-planning effort, birth rates fell by 43 per cent between 1972 and 1982. Where the effort was moderate, they fell by only 23 per cent. Where effort was weak or non-existent, they dropped by only 11 and 3 per cent.[49]

Although people can and do plan their families without modern contraceptives, programmes have an impact for two reasons. First, there are large numbers of women who have more children than they want. The share of unwanted births is 10 to 20 per cent. Among women who want no more children, between 43 per cent and 77 per cent are not using contraceptives. Making family planning available soaks up this unmet demand. Second, and more crucially, family planning creates demand for more family planning. The technology, with the education and discussion that accompany it, make more women aware that they can plan their families – at the same time as providing easier ways of doing so. It acts as the focus for a cultural revolution, and speeds that revolution along.[50]

The use of contraception has increased very markedly in developing countries. In 1960–5 only 9 per cent of fertile women were contracepting. By 1983 45 per cent were. But performance was still very uneven. In East Asia, where birth rates fell fastest, three out of four women were contracepting in 1983, and in Latin America 56 per cent. But in South Asia only one women in three was using contraception, and in Africa less than one in seven.[51]

The key to success in family planning is improved access. Contraceptives must be made available, within easy reach of every home, at low prices affordable by the poorest, and in user-friendly ways. A wide choice of methods, to suit different stages of life, will enlist more couples. The right to a choice of method is important, as well as the right to choose family planning. If couples are offered only one method – especially an irreversible one like sterilization – they will often opt for no method, or for abortion.

Again, there are wide differences in accessibility between regions. Some 95 per cent of people in East Asia had good access to contraceptives in 1982 – that is, they could get hold of supplies without spending more than two hours per month, or more than 1 per cent of their wages. The proportion with good access was 57 per cent in South East Asia and Latin America, and 54 per cent in South Asia. But in the Near East and North Africa the proportion fell to 13–25 per cent. In sub-Saharan Africa it was only 9 per cent.[52]

Equally important, but harder to measure, is the status of women. Many of the most rapid reductions in birth rates have occurred in countries where women have the right to inherit land – either by tradition, as in Kerala, Thailand, or because of revolution, as in Cuba or China.

Pulling out all the stops

All the factors that reduce population growth rates interact. Lowering infant mortality lowers the birth rate. But infant mortality can be cut dramatically by having fewer more widely spaced children, and avoiding pregnancies below the age of nineteen or above age thirty-five. Longer breast-feeding, which cuts infant deaths, also spaces children more widely. Better female education means fewer births per woman, and also better health for children – children of women with at least seven years of education have a 15 per cent better chance of surviving than those of women with no schooling. Better health improves education, since fewer days are missed through illness. Children from smaller families tend to be better nourished and healthier, and do better at school.[53]

Any progress in education, child health, women's status or family planning helps to slow population growth. But progress on all fronts together has a far greater impact. The effect is multiplied, not added.

What is needed, then, is a broad programme of developing

health, education, women's status and family planning, with priority to reaching the widest possible numbers.

Giving priority to human resource development widens options for future generations by widening options for the present. It combines equity with conservation. It reduces absolute poverty and improves the quality of life for everyone. It meets the basic human rights to health, education and free choice of family size. It improves women's status and prospects.

Two special priorities stand out. Of the expected 6 billion increase in future human numbers, 4.3 billion will occur in today's poorest countries. Regionally, 4.1 billion will occur in Africa and South Asia. A focus on these two regions, and other low income countries, will have a massive impact, reducing future world population, and improving the quality of life where it is lowest.[54]

The future impact of AIDS in these regions could be harsh. In the medium term, it will cut population growth below what is expected. But even the most Machiavellian should find no grain of relief in this. AIDS will kill men and women in their prime, leaving orphans and the aged. It will deepen poverty and cut resources available for human resource development. Unless its spread is halted by education and provision of condoms, AIDS is more likely to delay the shift to lower fertility than to speed it.

The second priority is women. Their position is pivotal. The number of children women choose to have is lower when they are more educated, have equal rights to property and equal access to work with equal pay. A free choice of family size, by women free to determine other aspects of their destinies, would improve the prospects of today's women, of future generations of men and women, and of other species. But women's rights are not always freely conceded. They often have to be fought for, and women who are literate and educated are better prepared for that struggle. Of all the aspects of human resource development and the shift to a sustainable world, improving female education is the crux.

Human resource development is not an alternative to pushing for economic growth. It is, when combined with freer markets, the best means of doing so. Those countries that have seen the most sustained economic growth since 1965 have tended to be

the ones with slower population growth. Many of those with the
fastest population growth have seen stagnation or even decline
in per capita incomes. In a sample of 79 developing countries,
population growth explained 40 per cent of the variations in
growth in incomes per person during the 1980s.[55]

Finally, by slowing population growth, human resource
development reduces environmental damage.

No other strategy offers such a combination of rich payoffs.

Working separately on population, consumption or technology will reduce the rate of environmental damage.

But our basic equation for environmental impact (p. 240) is a
multiplication sum. Progress or regress are compounded. A
doubling of population, of consumption, and of environmental
impact per unit of consumption means an eightfold rise in
damage to the environment. Equally, a halving of all three
would cut damage to one-eighth of its former level.

The three elements interact, just as the components of slower
population growth do. But not in the same way. Slower
population growth is broadly associated with increased consumption per person. Rates of deforestation may be lower, but
pollution may increase as incomes rise and industrialization
proceeds. If we don't work on all three elements, progress in
one area could be wiped out by regress in others. If we do work
on all three, progress in one area multiplies progress in the
others.

And remember that these three factors are only our
proximate determinants of environmental impact. To reduce
their impact also means working on the whole range of factors
reviewed in Chapter 18, from abolishing poverty and reducing
inequality, to improving markets, economic policies, democratic
rights, and institutions for controlling commons.

The task of balancing population, consumption and technology with the environment will be Herculean. Unless we pull out
all the stops, the damage incurred before we stabilize our
relationship with the environment will be catastrophic.

20

IF IT BE NOT NOW, YET IT WILL COME:
towards the third revolution

> I do not know
> Why I yet live to say 'This thing's to do,'
> Sith I have cause, and will, and strength, and means
> To do't.
>
> Shakespeare, *Hamlet*

> When I walk by the peasants' woods which I have
> saved from cutting down, or when I hear the rustling
> of the young copse planted by my own hands, I
> realize that the climate is to some extent in my own
> power, and that if in a thousand years human beings
> are to be happy, I too shall have had some small
> hand in it.
>
> Anton Chekhov, *Uncle Vanya*, 1897

Sophocles' tragedy *Oedipus Rex* has a claustrophobic inevitability. Oedipus' fate is sealed from the outset. Long before the play begins, he has unwittingly killed his father and married his mother. We witness only the slow dawning of the truth and Oedipus' self-destruction.

But the tragedy of *Hamlet*, like all Shakespeare's great tragedies, is avoidable. Hamlet knows from the outset that something is rotten in the state of Denmark. His uncle Claudius has married his widowed mother. After seeing his father's ghost, Hamlet knows what is wrong: Claudius murdered Hamlet's father.

But Hamlet doesn't act on that information. He wants to give evil a fair chance to prove itself innocent. He sets up the charade of the play within the play. When Claudius sees his crime re-enacted and starts in terror, Hamlet knows beyond any doubt that Claudius must be killed.

He comes upon the usurper at prayer, and toys with the idea of killing him there and then. But he passes up the opportunity. Claudius, who has seen Hamlet's game, sends him forthwith to England, to be executed on arrival.

Thereafter Polonius, Rosencrantz, Guildenstern and Ophelia die.

Hamlet escapes and returns to Denmark. Though Claudius has made an attempt on Hamlet's own life, still Hamlet delays. He plays into Claudius's trap, is slashed by Laertes's poisoned sword, and watches as Laertes and his mother die, poisoned.

Six innocents have been killed. Deadly poison is in Hamlet's own blood. He has less than half an hour to live.

Only at this ultimate extremity, when he is left with no choice and no respite, does Hamlet finally kill Claudius.

The Hamlet syndrome

The human race is not like Oedipus. We are not inescapably doomed, whatever we do. But we are very much like Hamlet. We may be doomed, if we carry on procrastinating.

In almost every case we have let environmental problems reach the stage of crisis before doing anything about them. Blue and humpback whales were hunted close to extinction before whaling was banned. Forests had to start dying before acid rain – first raised as a problem in 1872 – was taken seriously. An ozone hole of 14 million square kilometres had to appear to galvanize international action on chlorofluorocarbons.

The lessons of our past are not encouraging. They seem to suggest that catastrophic damage must occur, or obviously impend, before we are shaken into taking decisive action.

Costs and damages are always higher if action is taken only when it is forced. Hamlet's delay cost seven innocent lives, including his own. Our delays cost too, in species and soil lost forever, toxic waste dumps and nuclear refuse left as a legacy to our grandchildren.

Is this how it will always be?

Will global temperatures have to rise by three or four degrees to make us act on global warming? Will half our forests have to die before we cut our use of private cars? How many toxic red tides must occur before chemical fertilizer use is controlled? Does desertification have to eat up a third of a village's lands before the village will act?

Will we stagger on like this for ever, acting only when we have to, diminishing the future's options, even risking the absence of a future?

Can we shake the Hamlet syndrome? Can we learn to act before we are forced?

It seems almost an inexorable law of history that we only respond to environmental crisis when we have to. But history has another, less pessimistic lesson. It is that there are no inexorable laws of history. The rise of communism was not, as Marx and Lenin claimed, the inevitable result of iron laws. It was the outcome of determined action that succeeded against all probabilities. Nor was the collapse of communism the result of any contradicting set of laws. The system could have teetered on for decades, but for Gorbachev's bold break with seventy years of history.

The exercise of free will has altered history. Human determination can transcend determinism.

Alternative futures

In the last chapter we looked at population projections for the next century and a half. I gave the medium figures, because these represent what demographers believe is the most likely outcome.

But forecasting is an uncertain art. The only certainty is that over the decade up to the end of the millennium, world population will increase by between 89 million and 112 million each year. The difference between these low and high United Nations' projections is not large, because the new parents of the year 2000 were already between six and sixteen years old in 1990.[1]

After that the future is wide open.

By 2010 we could number as few as 6·8 billion, if every country takes the task of human resource development seriously. Or as many as 7·6 billion, if they do not. In 2025, world population could be anything between 7·6 billion and 9·4 billion. By 2050 we could be 7·8 bilion – or 12·5 billion.

What happens in the rest of the twenty-first century varies even more widely depending on government policies and individual decisions.

The UN's medium projection assumes that female fertility in developing countries will fall towards replacement level, just over two children per woman. In developed countries, and even in developing countries in East Asia and the Caribbean, women in 1990 were already having fewer children than were needed to replace themselves. The medium projection assumes that fertility in such countries will rise back to two again. In that case world population would stabilize at 11·5 billion.

It is only since the early seventies that any country has fallen below replacement level fertility. There is simply no precedent to show what happens next. If governments grow nervous about declining populations and introduce incentives for childbearing – or if they allow large-scale immigration – then fertility could rise to two again. But if they allow individual decisions to take their natural course, then fertility could remain low, and population would start to decline. On the low projection, Western Europe's population will peak in the year 2000. By 2025 it will have fallen by 8 per cent.[2]

Suppose that developing countries make the most strenuous efforts in education, health, family planning and women's status from now on. Suppose, too, that the number of children each woman has falls below two and comes to rest there. In that case the world's populaton would peak in the middle of the next century and start to fall after that.

If the total fertility rate stablized at 1·96 children per woman, the world's population would fall to only 5·6 billion by the year 2150. If the rate stabilized at 1·7 per woman, then world population in 2150 would be only 4·3 billion – about the same as in 1978. Such low fertility rates are quite possible. The current average for Europe as a whole is 1·7 children per woman. Eleven developed countries have even lower rates, ranging down to 1·42 in Denmark.

But we cannot be sure that fertility will stabilize at replacement level or less. In an insecure world, with ethnic and national conflicts rife, it might stick at a somewhat higher level.

If it came to rest at 2·17 children per woman, then world population in 2150 would be 20·7 billion. If it lodged at 2·5, world population would be 28 billion – over five times its current level. In both these cases growth would continue at a steady pace thereafter.

And of course fertility might not stabilize at all. It might oscillate.

Blips, slopes and climbs

The prospects are exhilaratingly uncertain.

In the long term, the present human population explosion could turn out to be a mere blip in earth's history, lasting perhaps no more than 200 years. We would rise to a peak, probably not more than double our present numbers. Then we would decline. If we got too few for comfort, we could institute incentives for childbearing to keep at a stable level.

This 'blip' scenario is more encouraging than the bleak destinies that some have foreseen for us. It would provide more freedom of action for each one of us, and more chance for other species to survive. But it gives no cause for complacency. For during that blip we could cause a devastating amount of damage to bio-diversity and to the earth's climate – damage which might well be irreversible. Even if population growth is only a blip, it is in our interests to ensure that the peak is as low as possible, to keep damage to a minimum.

Or: our present expansion might turn out to be the steepest middle section of an S-shaped slope, leading to a rather inhospitable plateau of 10–14 billion. Such a world would involve very serious loss of original forest, at least 20 per cent loss of species, restrictions on use of water in most countries, constraints on many types of pollutant, and a high proportion of income spent on preventing further damage to the environment.

If international or ethnic tensions encouraged continued population growth, then we may be in for a longer climb. Short of colonizing space – the energy costs of which would further damage our earth base – we would still hit a ceiling some way further on, probably before we reached 30 billion or so. Assuming such numbers were sustainable, life could come to resemble the prophetic film *Soylent Green*, where meat and soap are exquisite rarities; nature is glimpsed only in videos played to euthanasia volunteers as they die; and human corpses are recycled as high-protein biscuits.

Even the most optimistic of these scenarios is fraught with danger. In all of them, the lower the peak population, the less will be the irreversible damage, the fewer will be the restrictions on our freedom over the next century and thereafter.

Everything depends on the decisions of each one of us, and of the governments that create the context in which individual decisions are taken. If we launch now on the sort of combined strategy outlined in the last chapter, there is a fair chance that we could pull through without catastrophic damage.

Everything is still to play for.

That we would do, we should do when we would

It is often said that the 1990s will be the decade that decides the fate of the planet. That may be so. More probably, it is the second in a run of three or four decisive decades.

We wasted the first. The 1980s were our best chance to act. We passed it up. Debt made the developing countries into net donors to the rich countries. Social spending was cut back in Africa and Latin America. Progress in education and health halted and reversed. Fertility did not decline as expected. The cold war ended, but the peace dividend seemed likely to go towards cutting taxes, or helping former communist countries, rather than to helping the Third World in any significant way.

In the 1980s we came upon Claudius at prayer – and we let him live to do more evil.

The contrast between Claudius and Hamlet is in many respects the core of Shakespeare's play. Where Hamlet is wavering, Claudius is decisive. He is not made coward by conscience, nor disabled by thinking too precisely on the event. For him there is no chink between thought and deed. As soon as he senses danger he sends Hamlet to England. As soon as

Hamlet returns, Claudius devises not one, but two plots to kill him.

We must learn to act swiftly like Claudius. We must break the fatal habit of acting too late.

To do so means working to reduce population growth and excess consumption, and to change damaging technologies for benign. It also means changing all those indirect factors that affect the three primary agents. We must strengthen our institutions, to shorten the delays in perceiving and acting on environmental problems (see Chapter 17). We must improve our capacity to monitor changes in the environment. We possess the technology to do so: what we need is the commitment of funds and manpower. We must educate farmers, foresters and fishers to spot early warning signals of damage, so they will become concerned at an earlier stage.

We must spread democracy where it does not exist. Where it does, we must strengthen it by improving education, and creating the right of access to government and company information on environmental impacts. We must introduce free markets where they do not exist. And where they do, their blind spots should be removed: environmental and other social costs and benefits must be fully accounted.

We must increase security of tenure or ownership over land. And for forests, rangelands, rivers, oceans and atmosphere, we must strengthen community, national and international control.

We must improve our understanding of environmental processes. We are really only beginners in understanding the incredible complexity of ecosystems – including the biggest ecosystem of all, the planet Earth. This means devoting a great deal more resources to environmental sciences. And it means creating the framework for a much more interdisciplinary approach to research and to first degrees. The demands of the environment will break down the compartmentalization of knowledge. This process is already visible in research pro-grammes on planetary change, which could well become the master-science overarching all others. As part of this we desperately need an overarching science of human interactions with the environment, combining demography, socioeconomic and technological studies with dynamic analysis of the physical environment.

Finding and spreading appropriate technologies is the final stage of the process. We must give far higher priority to research into environmentally benign technologies, especially in energy production and in agriculture.

Aspects of resilience

Life is more resilient than most people realize. It bubbles out of hot fissures in the ocean bed. It reddens polar snowfields. Life is hard to keep down. Desertified rangelands recover when rested. Logged forests regrow if they are left alone. Even human disturbance can create a mosaic of cleared areas, regenerating areas, wasteland, field, field edge. Some sites in Israel, after ten millennia of severe human disturbance, have a species diversity almost as high as moist tropical forest.[3]

Human culture is resilient too. Our dominance on earth is not due to rigidity, but to our speed in adapting to changes in the environment. In every sphere we have seen adjustment processes at work. When population growth, consumption changes or technology create problems, people work to overcome those problems – though they do not always succeed.

There may also be a higher-level adjustment process, working at the level of whole economic and value systems.

Hunter-gatherers have an almost religious reverence for nature. Animals, plants, even rocks and rivers, have souls. Plants and animals are taken as needed – but with due concern for sustainability, since overconsumption of wild food sources will deplete them and threaten group survival.

However, most early hunter-gatherers did eventually consume beyond the limits of their wild food supply. As populations grew, agriculture spread, and allowed population to grow further. With agriculture comes a shift in attitudes. Wild nature is no longer the main source of food, but the cradle of weeds and pests and the den of predators. And it is potential land to be cleared for farming. Attitudes to wilderness change from reverence to revulsion, fear, desire to exterminate, tame and convert.

These negative views persisted up to the mid eighteenth century in Britain. The Alps were considered 'hideous' in 1621 by traveller James Howell. In 1681 John Houghton denounced Hampstead Heath as a 'barren wilderness' in urgent need of cultivation. To Dr Johnson the Scottish Highlands were a 'wide extent of hopeless sterility'.[4]

Absence makes the heart grow fonder

It took complete separation from nature to bring back reverence for nature. Large scale urbanization, combining physical crowding with social isolation, drove the romantics to seek out the solace of unspoiled countryside – just as it had earlier driven the intelligentsia of imperial Rome to idealize the rustic simplicity of their ancestors.

To Wordsworth, London in 1791 was a 'huge fermenting mass of humankind', a 'perpetual whirl of trivial objects'. In lonely rooms, amid the din of towns and cities, he consoled himself with memories of 'wild secluded scenes' and called himself a nature worshiper. 'Nature then,' he wrote, 'to me was all in all.'[5]

As urbanization progressed, there was a reversal of earlier ideals of beauty. Gardens, once rigorously geometrical, became deliberately unkempt. In 1822 the painter Constable wrote of his aversion to landscaped parks: 'It is not beauty because it is not nature.'[6]

In more recent times further separations from nature have occurred. Chemicals replaced manure and compost on farmland. Processed foods supplanted raw. Synthetic fibres ousted natural ones. Fossil fuels took the place of wood and animal power. Plastic elbowed out cardboard. Cement and steel superseded timber and clay. In general, the artificial dislodged the natural. The inevitable reaction to this process was the rise of organic farming, of raw and unprocessed foods, of handicrafts.

High levels of industrialization and urbanization lead to a plateau in human relations with the environment. There is a high density of waste, of pollution, of traffic congestion. At the same time democracy is relatively developed. Pressures to control pollution are pronounced, and heeded. Leisure and wide car ownership provide easy access to countryside. Pressures for conservation grow. An increasing proportion of land is set aside as national parks and nature reserves.

But like the lower-level adjustments, this macro-adjustment process is too slow to prevent large-scale damage to the environment: indeed it comes about only where damage has already occurred. In the most advanced developed countries it is beginning to reach an active phase, where controls are applied systematically and progressively on industrial and farming processes. But there is still a marked hesitancy in tackling pollution from individual domestic sources – since this could lose votes. In no country on earth has the adjustment come anywhere near the stage where environmental damage has ceased.

Most developing countries are in transition between phases two and three, agriculture and industrialization. At this stage the level of damage is truly massive. There is continued forest clearance for agriculture, reckless logging, industrialization without effective democratic controls on pollution.

Towards the third revolution.

The agricultural and industrial revolutions were the response to pressures of population growth on the environment. Shortage of wild food resources led to the first, shortage of wood to the second. In the first case we adjusted technology and consumption (reducing the variety of our diet); but population growth increased. In the second we adjusted technology, shifting to fossil energy; but population growth accelerated further. Later, in response to the costs and opportunities of industrialization and urbanization, we finally began to adjust our populations.

We are now in the throes of the third revolution. This time the spur is not resource shortage, but the impact of waste and wasting. It will demand responses right across the board, in population, consumption and technology and everything that affects them. It will continue until we reach a sustainable balance with our natural environment.

There is one factor that could short-cut the adjustment process: it is the major shift in values that has begun. Such shifts are determined by historical forces. But once in motion on a large scale the new values motivate human action and themselves become historical forces. Witness the rise of Christianity or Islam, the democratic revolutions of 1776–1848, or the socialist revolutions of 1917–75.

The current value shift is possibly the most far-reaching in our attitude to nature since the rise of transcendental religions between 600 BC and AD 700. The new values have already gestated, in developed countries and among the intelligentsia of more urbanized Third World countries.

They are now spreading with the speed of a new religion. And the global nature of our environmental crisis – like the threat of hell fire in early Christianity – is the most persuasive of its evangelists.

Indeed the new values have a quasi-religious content. There is a tendency away from the concept of a God outside or above the universe, towards a divinity inherent in everything. Away from transcendentalism, towards immanence and pantheism. There is a parallel shift away from a simple mechanical and determistic view of the physical world, towards a more holistic view. The world is not an engine, but an organism. Gaia is treated as a deity by some.

The concept that humans by right are masters of the earth, the pinnacle of creation, is already giving way to the idea that we are stewards, with a responsibility to look after the patrimony with which we have been entrusted. This idea in turn phases into the view that we are members of a community. If one member overreaches itself, the others rebel. If we damage nature, we damage ourselves. We cannot, try as we will, control nature over the long run: we can only co-operate with it.[7]

The new philosophy involves a new ethics of extended altruism. Altruism has prevailed within the primary group since hunter-gatherer times. The great universal religions extended altruism to cover whole societies. In the modern world it is applied in theory – though not in practice – across national boundaries. The goal of equity, social justice and the abolition of absolute poverty within the present generation has wide influence among individuals, if not among governments.

The next ethical principle is altruism towards future generations. We must not prosper at the expense of poverty and constraint for our descendants. We ought to leave them a world that is not diminished in its richness. Development must be sustainable, or it is no more than a brief interlude between aeons of human misery.

The third principle is altruism towards other species. They have rights as well as humans. We should not thrive on earth by exterminating other organisms. We can hardly avoid confining them to smaller areas. But at least there we should work to preserve them and the ecosystems of which they are part. This concept is also inherent in the second. If we don't treat other species equitably, we impoverish the world that our children inherit.[8]

The new philosophy and the new ethics are in some respects not new at all. They are the resurrected wisdom of the hunter-gatherer, extended out from the local habitat to the entire planet. In the same way the new forestry and the new agriculture will emphasize the diversity that hunter-gatherers prized. Human attitudes to the environment have come full circle.

The readiness is all

Adjustment will occur. Sooner, with less damage, if we are wise. Later, with more damage, if we are not. Hamlet had to kill Claudius: the only question was when, and whether Claudius would kill Hamlet first. If Hamlet had acted when he should, seven innocent lives would have been spared.

Much of the last chapter was aimed at policy makers. But environmental problems are the outcome of individual actions multiplied millions of times. That is one reason why they are so intractable, but at the same time it is a source of hope. Many changes in environmental behaviour do not have to pass through the whole political system to become effective. Of course political action helps: not just campaigning for wider changes, but for local ones. Anyone can change their own behaviour straight away, as an individual, as a member of a family, of a local community and a workplace. We don't have to wait till collective disaster forces us to change on a collective scale.

We can plant trees. Turn our garden into a wildlife retreat.

Pay for others to conserve habitats and restore degraded ones – nature will need millions of midwives over the coming decades. If we are farmers or foresters, we can try to maintain production, or long-term income, in ways that preserve or enhance biological capital rather than reducing it.

Remember the waste monuments. Do we really wish to be commemorated by a garbage mausoleum one to four thousand times our bodyweight? Or by 240-tonne carbon balloons, three and a half thousand times our weight?[9]

We can turn lights off when we leave a room, starting now. Lower heating or air conditioning controls. Insulate the home. Walk, cycle or use stairs where we can. Wash hands in cold water. Next time we renew a light bulb or an appliance, make sure it is the most energy efficient we can afford. Choose a car for maximum fuel economy, not maximum power to impress.

Use consumer power to change what manufacturers provide. Eat organic, fresh and unprocessed foods if you can afford them. Eat less red meat. Recycle what you can. Refuse unnecessary packaging. Don't buy canned drinks. Buy recycled goods. Use things till they wear out. Repair anything that can be repaired. Don't trash anything that someone else could make use of. Have no more than two kids.

Try to break the habit of impressing other people through conspicuous consumption. Status should be determined by how little we damage the earth, or how much we enhance it – not by how much we contribute to its spoliation. Excess consumption should become a matter of scorn and shame, not pride.

The list is endless, and we can add to it each day. Every new addition is a new victory. For it is a war – one that will go on until we finally achieve a sustainable balance with our environment.

It is not only the Kalsakas and the Hatias that will suffer if we procrastinate. It will be our own children and grandchildren. The time is near when every child will ask its parent 'What did you do in the environment war, mum and dad? Were you one of those who helped to destroy my future? Or were you one of those who helped to save it?'

Remember, if your children are still at home, that they will still be alive in the year 2040 or 2050. Remember that your grandchildren will see the world of 2070 or 2080.

The future is really that close. And the future will judge us harshly if we do not change. For we risk being remembered as the generation which, like the Kwakiutl chiefs, sent for more possessions to heap on the potlatch fire while the earth burned. The generation that in the space of only three or four decades closed up the horizons, hemmed in our children's freedom, left them a duller, uglier, poorer world.

The generation that caught Claudius at prayer, and let him go on living.

The time is out of joint. Like Hamlet, we were born to set it right.

Should we succeed, the present age will be seen as the age of the Third Revolution. It will not, any more than the first two, be immediate. It began its infancy no more than two decades ago. In human terms, it is now an inexperienced and impulsive youth. Early adulthood is at least a decade away, maturity twenty years ahead, full fruition more distant still.

It will be known as the age when human numbers, consumption and technologies were shifted into sustainable balance with the environment. When we developed social arrangements to keep us in balance, despite inevitable and permanent change. When the needs of living humans, of future humans and of other species were reconciled.

In all of human history, no generation has ever borne such a responsibility on its shoulders.

APPENDIX
Assessing population impact on environment

Most of the debate on the impact of population growth on the environment, or lack of it, has been conducted in an unscientific manner, in a vacuum of evidence, deprived of the oxygen of hard data.

But things become much clearer when they are looked at quantitatively and in terms of physical mechanisms. Then population tends to take its proper place as one of the key direct factors.

Paul Ehrlich's formula:

1. impact = population × affluence × technology

is often used to discuss the connections. It has the appearance of mathematical rigour, but it does not help towards greater precision. The term 'affluence' demands a value judgement about what is 'affluent' and what is not. It implies that subsistence consumption has no impact on the environment, when in fact all levels of consumption, even of hunter-gatherers, have some impact.

For more precision, 'affluence' needs replacing with 'consumption per person'. This covers everything from eating wild roots or building a palm leaf shelter, to walling your bathroom in marble or driving a Porsche at 200 kilometres an hour down the *autobahn*. The term 'technology', too, is not capable of measurement as such and needs replacing, with 'environmental impact per unit of consumption'.

Thus we have:

2. environmental impact =
population × consumption per person × impact per unit of
consumption

Impact can refer to any one of our three main forms of
interaction with the environment: our use of primary resources
from minerals to water and land; our physical occupation of
space; or our output of pollutants.

The formula, which is a rephrased version of Barry
Commoner's (see p. 239) can be used in many different
situations. It can be applied by, or to, high-consuming
individuals, to give them some idea of the environmental cost of
their lifestyles. It helps to make tangible what so often is
invisible. It is identical to the formula for farm area on
pp. 103–4.

It can also be applied at a particular moment in time to
countries or regions. The formula makes it clear that, for any
given level of consumption per person or technology, higher
population means higher environmental impact. Fewer people
mean lower total consumption and waste.

To assess what contribution population makes to increases in
environmental damage – to try to assign relative blame – we
have to look at changes over time. What we need is some
measure of the share of each of our three elements in overall
impact. If all three are pushing upwards, we can simply express
the change in each one in turn as a percentage of the total
change. So here:

3. population impact = $\dfrac{\text{annual \% change in population} \times 100}{\substack{\text{annual \% change in use of resource or} \\ \text{output of pollutant}}}$

Sometimes one or more of the three factors may be tending to
reduce environmental impact. We can distinguish between
upward and downward pressures by 'scoring' the upward
pressures out of +100 per cent, and the downward pressures
out of −100 per cent.

It might help at this point to look at some concrete examples.
Take the case of the expansion of agricultural land. This is not
merely physical occupation of space and use of the land
resource. It is a prime source of environmental damage, cutting
down forests, draining wetlands, ploughing up natural grassland,

replacing natural ecosystems of high species diversity with artificial ones of low diversity.

Let us see if we can discover what share in the increase in farmland we can attribute to population growth (see Table 1, page 309). Between 1961 and 1985, population expanded by 2·3 per cent a year in developing countries. Agricultural production rose by 3·3 per cent. So production per person, which we can take as our consumption factor, rose by 0·9 per cent. Farmland, meanwhile, expanded by only 0·6 per cent a year. The farmland used per unit of agricultural production – the impact per unit of consumption – actually declined by 2·6 per cent annually, because yields were increasing. Technology in this case exerted a downward pressure on environmental impact, at least in terms of land used. Only population and consumption exerted upward pressure.

In this case we can say that population growth accounted for +72 per cent of the growth of farmland, and increase in consumption per person for +28 per cent. Technology gets a score of −100 per cent, since it was the only one of the three factors pushing towards lower use of farmland.

In developed countries, with slower population growth and faster increase in agricultural production per person, the contributions work out quite differently: here population growth accounted for only 46 per cent of the expansion in farmland, increased consumption per person for 54 per cent. Technology again scores −100 per cent of the downward factors.

Another example. Increase in livestock numbers is a factor in several types of environmental damage. Goats eat tree seedlings. Cows compact soil and vent methane. So let's do the sum again for developing countries between 1961 and 1985 (see Table 1, page 309). Meat and milk consumption grew by 3·2 per cent a year overall, per capita consumption (given the 2·3 per cent growth in population) by about 1 per cent. The number of livestock, however, rose more slowly than meat and milk production, by only 1·3 per cent a year. So the number of livestock required for each unit of meat and milk dropped by 2 per cent a year, due to technology changes such as better veterinary treatment, better breeds of animal and so on. Once again, technology was responsible for a net downward pressure on livestock numbers and so scores −100 per cent. Population

Table 1 LAND AND LIVESTOCK

FARM AREA	Annual % change 1961–1985				Impacts: % of change		
	Farm area	P	C	T	P	C	T
Developing	0·57	2·3	0·9	−2·6	**+72**	+28	−100
Developed	0·13	0·9	1·0	−1·8	**+46**	+54	−100
LIVESTOCK	Numbers of livestock	P	C	T	P	C	T
Developing	1·3	2·3	1·0	−2·0	**+69**	+31	−100
Developed	0·3	0·9	0·6	−1·2	**+59**	+41	−100

Notes:
Farm area: area of arable and permanent crops.
P = population increase
C = change in agricultural production per person
T = change in farm area per unit of agricultural production. This is the inverse of yield.
Livestock: includes total numbers of cattle, sheep and goats.
P = population increase
C = change in meat and milk consumption per person
T = change in number of livestock needed for each unit of meat and milk consumption.

Data source (including population): Food and Agriculture Organization, *Country Tables 1990*, FAO, Rome, 1990, pp. 320–1 and 334–5.

accounted for 69 per cent of the upward pressure on numbers, and consumption for the remaining 31 per cent.

Increasing fertilizer use alters soil composition, contributes to emissions of nitrous oxide which stokes the greenhouse effect, and pollutes rivers, lakes and coastlines. Here we are dealing with a massive and rapid technological change in agriculture. The population impact is correspondingly lower (see Table 2, page 310). It accounted for only 22 per cent of the increase in developing countries. Increased consumption – measured by the growth in agricultural production per person – accounted for only 8 per cent. The technology factor – increased fertilizer per unit of agricultural production – accounted for 70 per cent of the increase.

In many cases, as Commoner points out, we don't know the

Table 2 FERTILIZER USE

REGI	Annual change % 1961–88				Impacts (% of change)		
	F	P	C	T	P	C	T
Africa	7·7	2·9	−0·7	5·4	**34·7**	−100·0	65·3
Latin America	8·4	2·4	0·1	5·6	**29·8**	1·8	68·4
Near East	10	2·8	0·1	6·6	**29·4**	1·2	69·4
Far East	11·1	2·3	0·6	7·9	**21·5**	5·5	73·1
Asian CPE	13·2	1·9	2·4	8·5	**14·9**	18·9	66·2
Developing Countries	11	2·3	0·9	7·5	**21·6**	8·1	70·2
Developed Countries	4·3	0·9	0·7	2·6	**20·7**	17·7	61·5
World	5·9	1·9	0·5	3·4	[32·7]*	8·7	58·5

Global population impact:

Weighted by 1988 populations: **22·1** per cent.

Weighted by 1988 fertilizer use: **20·8** per cent.

Notes:
F = Increase in fertilizer use.
P = Population increase
C = Increase in agricultural production per person.
T = Increase in fertilizer use per unit of agricultural production.

Data source: Food and Agriculture Organization, *Fertilizer Yearbook 1989*, FAO, Rome, 1990; 1961 data from FAO, *Country Tables 1990*, FAO, Rome, 1990.

* Square brackets in world results indicate unweighted figure, for illustration only. Only the weighted figures should be quoted in giving the population impact.

actual damage attributable to each unit of consumption. So we end up measuring the output of pollutants per unit of consumption. Take the output of air pollutants linked with energy consumption in OECD countries (see Table 3, page 311). In this case I have taken final consumption expenditure as the consumption measure. The data here allow us to break down the technology factor into two components: change in energy efficiency – energy requirements per unit of final consumption; and change in the amount of pollutant emitted for each unit of energy used, which has more to do with the mix of energy sources and the technology of emission cleaning.

Table 3 AIR POLLUTANTS FROM ENERGY USE IN THE OECD 1970–88

	Annual % change 1970–88					Impacts: % of change			
	Emissions	P	C	Te	Tp	P%	C%	Te%	Tp%
SO_2	−2·6	0·8	2·4	−1·7	−4·0	+25	+75	−30	−70
Smoke	−4·0	0·8	2·4	−1·7	−5·4	+25	+75	−24	−76
NO_x	0·7	0·8	2·4	−1·7	−0·8	+25	+75	−69	−31
CO_2	0·8	0·8	2·4	−1·7	−0·7	+25	+75	−72	−28

Notes:
SO_2 – sulphur dioxide
Smoke – particulate matter
NO_x – nitrogen oxides
CO_2 – carbon dioxide
Emissions = change in emissions of gas
P = population
C = change in private final consumer expenditure per person
Te = Energy technology – energy requirement per unit of consumer expenditure
Tp = Pollution technology – emissions per unit of energy requirements

All figures are percentages. The first five colums represent the annual percentage change over the 1970–88 period. The last four columns represent the percentage contributions – positive (upward) or negative (downward) of each factor to the overall change in emissions.

Data source (including population):
Organization for Economic Cooperation and Development, *State of the Environment 1991*, pp. 236, 241, 243, OECD, Paris, 1991.

Using the same methods, we find that even in the slow growing OECD countries, population was responsible for +25 per cent of the upward pressure on emission of air pollutants, and increased consumption for +75 per cent. But the results varied considerably between countries. In the USA, where population growth was maintained by immigration,, the population factor amounted to +30 per cent of the upward pressure on emissions. In West Germany, where population grew by only 1·3 per cent over the whole eighteen year period, the population impact was only +2·4 per cent.

Technology was a downward pressure. In the case of sulphur dioxide and smoke, total emissions actually fell, respectively by 2·6 per cent and 4 per cent per year. Technological change aimed specifically at reducing pollution per unit of energy consumed (shown as Tp in the Table 3) appears to have had about twice as much downward impact as the general increase in energy efficiency (shown as Te).

The results for two other pollutants, oxides of nitrogen and carbon dioxide, were quite different. These increased, by 0·7 and 0·8 per cent respectively, though not as fast as the increase in either population or consumption. So once again technology was a downward pressure, though much weaker than in the case of sulphur dioxide and smoke. But in these cases the impact of reduction in emissions per unit of energy used was two to two and a half times less than the increase in energy efficiency.

There are some cases where it is not feasible to measure separately the consumption and technology factors. For example, in the case of expansion of irrigation, it may be difficult or impossible to discover exactly how much foodstuff was produced on irrigated as opposed to non-irrigated land. In the case of chlorofluorocarbons, it is hard to isolate the consumption element, since this would mean aggregating incommensurables like refrigeration, air conditioning, packaging and solvents. There are other cases where consumption and technology are not easily separable: new types of product, for example, such as fridges, video recorders, or computer games, which are not merely providing an old need in a new way (as cars do for transport) but where the product itself creates the consumer 'need' which it supplies.

In many of these tricky cases it is still often possible to get some idea of the population contribution, using Formula 3 above (page 307). In cases where all three factors are pushing in the same direction, this formula will give the same results as Formula 2 (page 307).

When one or other of the three elements are declining, and consumption and technology measures are absent, one can still get some idea of the impact of population growth, compared with the other two elements, using the formula:

4. population impact =
    absolute value of population growth rate* × 100

---

absolute [population growth] + absolute [growth in pollutant
        − population growth]

* Absolute = ignoring whether the value is positive or negative.

Table 4 WORLD CARBON DIOXIDE EMISSIONS 1960–1988
From fossil fuels and cement only – short method (see page 312)

	Carbon dioxide (mill to C)			Population (millions)			
	1960	1988	% Increase per annum	1960	1988	% Increase per annum	Population impact %
Africa	40	170	5·3	279	605	2·8	**53**
Asia (dev)	297	1193	5·1	1574	2876	2·2	**43**
Latin America	82	267	4·3	218	430	2·5	**57**
North America	852	1430	1·9	199	272	1·1	**60**
Europe	732	1180	1·7	425	496	0·55	**32**
USSR	396	1086	3·7	214	284	1·0	**28**
Japan	64	270	5.3	94	122	0·95	**18**
Oceania	28	75	3·6	16	26	1·8	**49**
Developing Countries	418	1630	5·0	2072	3911	2·3	**46**
Developed Countries	2072	4040	2·4	948	1200	0·85	**35**
World	2490	5670	3·0	3020	5111	1·9	**[64]**

Global population impact:
Weighted by 1988 populations: **44%**
Weighted by 1988 emissions: **42%**
Weighted by increase in emissions: **41%**

Data source:
Emissions: United Nations Environment Programme, *Environmental Data Report 1991/ 92*, Blackwell, Oxford, pp. 16–20.
Populations: United Nations Population Division, *World Population Prospects 1990*, United Nations, New York, 1991.

This last formula, however, does not assess population's share of the change. Instead it measures the relative strength of the population influence compared to consumption-plus-technology considered together, regardless of whether the influences are positive or negative.

One other caution relates to the geographical level of analysis. Applying these formulae at global level can be misleading, and gives paradoxical results. Take, for example, increases in emissions of carbon dioxide from fossil fuel and cement making (see Table 4, p. 313). Between 1960 and 1988 these increased in developing countries by 5.3 per cent a year, while population increased by 2·3 per cent a year. The population impact is therefore 46 per cent (2·3 × 100 ÷ 5.3). In developed countries, emissions grew by 2·41 per cent a year and populations by 0·85 per cent. The population impact is 35 per cent.

At global level we find that emissions rose by 3 per cent, while population rose by 1·9 per cent. The population impact therefore appears to be 63 per cent – much higher than the average of the two separate measures. This happens because the slower growth in the much higher level of emissions in the North results in a lower global figure for emission growth.

Hence global figures for population impact cannot be produced on the basis of global change figures in the three elements. It means that, as a minimum, results should be worked out separately for developing and developed countries. Ideally regional and country level results are needed. We can still arrive at global figures for population impact, by averaging the separate results. The averages can be weighted in various ways. Weighting the regional results by 1988 populations, this gives the outcome that population growth accounted for 44 per cent of the increase in carbon dioxide emissions. Weighted by 1988 emissions, the population impact is 42 per cent. Weighted by total increases between 1960 and 1988, the population impact is 41 per cent.

There is no finality about these methods or results. The subject is young, and fluid, and alternative approaches may emerge. What I hope they do show is the scope for developing detailed quantitative studies in which population, consumption and technology impacts on the environment can be assessed.

NOTES AND SOURCES

Prologue: in the forest

1 This chapter is extracted from Paul Harrison, *Birdwing*, forthcoming.

1 The great debate

1 Ricklefs, Robert, *Ecology*, 3rd edn, W. H. Freeman, New York, 1990, pp. 320–1.

2 Highest births' figure from *Guinness Book of Records 1987*, Guinness Books, London, 1987, p. 12.

3 Population densities from Polgar, Stephen, ed., *Population, Ecology and Social Evolution*, Mouton, The Hague, 1975, pp. 29 and Table 1, p. 182; pre-1950 populations from McEvedy, Colin and Jones, Richard, *Atlas of World Population History*, Penguin, Harmondsworth, 1978, p. 342.

4 Post-1950 populations from United Nations Department of Economic and Social Affairs, *World Population Prospects 1990*, ST/ESA/SER.A/ 120, New York, 1991.

5 Plato, *Republic*, 460 *sq.*, and *Laws*, 740 *sq.*, Aristotle, *Politics*, 1325–6 and 1334.

6 Godwin, William, *Enquiry Concerning Political Justice*, ed. Isaac Kramnick, Penguin, Harmondsworth, 1976, pp. 767–77.

7 This version, and the much-modified and qualified second edition, are given in Himmerlfarb, Gertrude, ed., *On Population: Thomas Robert Malthus*, Random House, New York, 1960.

8 Ricardo, David, *The Principles of Political Economy and Taxation*, London, 1817.

9 Cited in Himmelfarb, *op. cit.*, p. xxvi.

10 Letter to J. B. Schweitzer, 24 January, 1865, in *Works*, vol. 2, p. 391; Critique of the Gotha Programme, *Works*, vol. 1, p. 29; Engels, Friedrick, letter to Albert Lange, 29 March, 1865.

11 Boserup, Ester, *The Conditions of Agricultural Growth*, Allen & Unwin, London, 1965. The theory is expanded and updated in *Population and Technology*, Basil Blackwell, Oxford, 1981.

12 Boserup, Ester, *The Conditions of Agricultural Growth*, Allen & Unwin, London, 1965, p. 118.

13 Ehrlich, Paul and Anne, *Population, Resources, Environment*, W. H. Freeman, San Francisco, 1970; *New Scientist* 36, p. 652.

14 Ehrlich, Paul and Anne, *The Population Bomb*, Ballantine, New York, 1969; *Reader's Digest*, February 1969; *New Scientist* 36, p. 655.

15 Meadows, Dennis *et al.*, *The Limits to Growth*, Potomac Associates, Washington DC, 1972, p. 170.

16 Ehrlich, Paul and Anne, *The Population Bomb*, Ballantine, New York, 1969.

17 Bentham, Jeremy, *An Introductory View of the Rationale of Evidence, Works*, ed. John Bowring, Edinburgh, 1843.

18 Simon, Julian, *The Ultimate Resource*, Princeton University Press, Princeton NJ, 1981; Simon, Julian, *Theory of Population and Economic Growth*, Basil Blackwell, Oxford, 1986.

19 See for example Lappé, Frances Moore, *Food First*, Souvenir Press, London, 1980; and Blaikie, Piers, *The Political Economy of Soil Erosion*, Longman, London, 1985.

20 Commoner, Barry, 'The Environmental Cost of Economic Growth', *Chemistry in Britain* 8(2), pp. 52–65.

2   Three billion years of environmental crisis

1 Lovelock, James, *Ages of Gaia*, Oxford University Press, Oxford, 1988, p. 33.

2 Ricklefs, Robert, *Ecology*, 3rd edn, W. H. Freeman, New York, 1990.

3 Lovelock, James, *Gaia* and *The Ages of Gaia*, both Oxford University Press, Oxford, 1979 and 1988.

4 The account of early crises is based on Cowen, Richard, *History of Life*, Basil Blackwell, Oxford, 1990; Lovelock, James, *Ages of Gaia*, Oxford University Press, Oxford, 1988; and Stanley, Steven, *Earth and Life through Time*, W. H. Freeman, New York, 1986.

5 The following section is based on Cowen, Richard, *The History of Life*, Blackwell Scientific Publications, Oxford, 1990; Lovelock, James, *The Ages of Gaia*, Oxford University Press, Oxford, 1988; and Beatty, J. Kelly and Chaikin, Andrew, *The New Solar System*, 3rd edn, Cambridge University Press, Cambridge, 1990.

6 Lee, R. B., cited in Cohen, Nathan, *The Food Crisis in Prehistory*, Yale University Press, New Haven, CT, 1977, pp. 28–30. Much of the following section is based on Cohen's book.

7 Origins of agriculture based on Redman, Charles L., *The Rise of Civilization*, W. H. Freeman, San Francisco, 1978; Wenke, Robert J.,

Patterns in Prehistory, Oxford University Press, Oxford, 1980; Cohen, Mark Nathan, *The Food Crisis in Prehistory*, Yale University Press, New Haven, CT, 1977; Boserup, Ester, *The Conditions of Agricultural Growth*, Allen & Unwin, London, 1965.

8   Harlan, Jack and Zohary, Daniel, 'Distribution of Wild Wheats and Barley', *Science* 153, pp. 1074–80, 1966.

9   Smith, Philip and Young, T. Cuyler, 'The Evolution of Early Agriculture and Culture in Greater Mesopotamia', in Spooner, B. J., ed., *Population Growth: Anthropological Implications*, MIT Press, Cambridge, MA, 1972.

10 Binford, Lewis, *An Archeological Perspective*, Seminar Press, New York, 1972.

11 Cohen, Nathan, *The Food Crisis in Prehistory*, Yale University Press, 1977.

12 Young, T. Cuyler, 'Population Densities and Early Mesopotamian Origins', in Ucko, P. J. *et al.*, eds, *Man, Settlement and Urbanism*, Duckworth, London, 1972.

13 Boserup, Ester, *The Conditions of Agricultural Growth*, Allen & Unwin, London, 1965; further refined and expanded in: *Population and Technology*, Basil Blackwell, Oxford, 1980.

14 Ruthenberg, Hans, *Farming Systems in the Tropics*, 3rd edn, Clarendon Press, Oxford, 1980.

15 This transition has been carefully studied for Africa by Pingali, Prabhu, Bigot, Yves and Binswanger, Hans, in *Agricultural Mechanization and the Evolution of Farming in Sub-Saharan Africa*, Report ARU 40, World Bank, Washington DC, 1985.

16 *Acts* x: 13.

17 See Coon, Carleton, *The Hunting Peoples*, Penguin Books, London, 1976; and Service, Elman R., *Primitive Social Organization*, Random House, New York, 1971.

18 See Harrison, Paul, *The History of Heaven*, forthcoming.

19 The following account is based on: Wilkinson, Richard G., *Poverty and Progress*, Methuen, London, 1973; Kellenbenz, Hermann, 'Technology in the Age of the Scientific Revolution', in Cipolla, Carlo, ed., *The Fontana Economic History of Europe*, vol. 2, Fontana, London, 1973; Lilley, Samuel, 'Technological Progress and the Industrial Revolution', and Sella, Domenico, 'European Industries 1500–1700', both in Cipolla, Carlo, ed., *The Fontana Economic History of Europe*, vol. 3, Fontana, London, 1973.

20 Kellenbenz, Hermann, 'Technology in the Age of the Scientific Revolution', and Sella, Domenico, 'European Industries 1500–1700', both in Cipolla, Carlo, ed., *The Fontana Economic History of Europe*, vol. 2, Fontana, London, 1973.

21 Lilley, Samuel, Technological Progress and the Industrial Revolution, in Cipolla, Carlo, ed., *The Fontana Economic History of Europe*, vol. 3, Fontana, London, 1973.
22 Wilkinson, Richard G., *Poverty and Progress*, Methuen, London, 1973.
23 *Ibid.*, p. 136.

3   The new limits to growth

1   Vogely, William, 'Nonfuel Minerals', in Repetto, Robert, ed., *The Global Possible*, Yale University Press, 1985, p. 458.
2   *Ibid.*
3   1950 figure from: MacKellar, F. R. and Vining, D. R., 'Natural Resource Scarcity', in Johnson, D. G. and Lee, Ronald, eds, *Population Growth and Economic Development*, University of Wisconsin Press, Madison, 1987; modern reserves from: United Nations Environment Programme, *Environmental Data Report*, UNEP, Blackwell Reference, Oxford, 1989, p. 416; and World Resources Institute, *World Resources 1990–91*, World Resources Institute, Washington DC, 1990, p. 145.
4   *Global Economic Prospects and the Developing Countries*, World Bank, Washington DC, 1991.
5   Organization for Economic Cooperation and Development, *The State of the Environment*, OECD, Paris, 1991, p. 226.
6   *Global Economic Prospects and the Developing Countries*, World Bank, Washington DC, 1991, pp. 20–1; Repetto, Robert, 'Population, Resources, Environment', *Population Bulletin* 42:2, 1987.
7   These rates are calculated on the basis of 1988 US consumption rates of 18·8 kilos of aluminium per person, 9·2 kilos of copper and 4·5 kilos of zinc, with world reserves of 3960 million tonnes of aluminium, 437 million tonnes of copper and 149 million tonnes of zinc. Figures taken from United Nations Environment Programme, *Environmental Data Report*, UNEP, Blackwell Reference, Oxford, 1989, Table 3.24 and World Resources Institute, *World Resources 1990–91*, World Resources Institute, Washington DC, 1990, Table 21.4.
8   Calculated on the basis of reserve figures in United Nations Environment Programme, *Environmental Data Report*, UNEP, Blackwell Reference, Oxford, 1989, p. 416; World Resources Institute, *World Resources 1990–91*, World Resources Institute, Washington DC, 1990, p. 145.
9   Calculated from Food and Agriculture Organization Economic and Social Policy Department, *Country Tables 1990*, FAO, Rome
10 Food and Agriculture Organization, *Food Outlook Statistical Supplement*, 1982 and 1990, FAO, Rome; prices are for US no. 2 hard winter

wheat, and US no. 2 Yellow maize. I have chosen years without major droughts. No two runs of years are strictly comparable – but there is no overall increase visible, even before allowing for inflation.

11 Food and Agriculture Organization, *Production Yearbook 1989*, FAO, Rome, 1990, Table 106.

12 *Ibid*

13 Poverty: World Bank, *World Development Report 1990*, World Bank, Washington DC, 1990, p. 29; malnutrition: Food and Agriculture Organization, *Fifth World Food Survey*, FAO, Rome, 1985, p. 25; calorie intake drop: Alexandratos, Nikos, *World Agriculture in the Next Century*, XXI International Conference of Agricultural Economics, Tokyo, 1991; food aid in 1988–9 was 10,043,000 tonnes of cereals (FAO, *Food Aid in Figures* 8(1), p. 36, FAO, Rome, 1990), enough to keep 40 million people alive for a year, on a survival diet of 250 kg per person.

14 Per capita food production since 1961 from: Food and Agriculture Organization, Economic and Social Policy Department, *Country Tables 1990*, FAO, Rome, 1990. Index recalculated using 1961 as base year.

15 Cereal production figures calculated from Food and Agriculture Organization, Economic and Social Policy Department, *Country Tables 1990*, FAO, Rome, 1990; sub-Saharan Africa cereals: Alexandratos, Nikos, *World Agriculture in the Next Century*, XXI International Conference of Agricultural Economics, Tokyo, 1991.

16 Food and Agriculture Organization, *Production Yearbook 1989*, FAO, Rome, 1990, Table 9.

17 Food and Agriculture Organization, Economic and Social Policy Department, *Country Tables 1990*, FAO, Rome, 1990.

18 *Ibid.*

19 *Ibid.*

20 Higgins, Graham *et al.*, *Potential Population Supporting Capacities of Lands in the Developing World*, FAO, Rome, 1982, pp. 35–7.

21 The figures on land reserves that follow are based on Harrison, Paul, *Land, Food and People*, FAO, Rome, 1984, p. 10 (for cultivable area); Food and Agriculture Organization, *Production Yearbook 1989*, FAO, Rome, 1990, for 1988 arable and forest areas; and Alexandratos, Nikos, ed., *World Agriculture: Toward 2000*, Belhaven Press, London 1988, Table A.7.

22 Developing country cereal yields from FAO data disks; Alexandratos, Nikos, ed., *World Agriculture: Toward 2000*, Belhaven Press, London, 1988, Table A.7. Yields for China and developed countries from Food and Agriculture Organization, *Production Yearbook 1988*, FAO, Rome, 1989. Plateau populations from Bulatao, Rodolfo *et al.*, *World Population Projections 1989–90*, World Bank, Washington DC, 1990.

23 Growth rates and target yields calculated from sources in note 13.

24 Fertilizer growth from Food and Agriculture Organization, *Fertilizer Yearbook 1989*, FAO, Rome, 1990. Cattle-to-people ratios calculated from Food and Agriculture Organization, Economic and Social Policy Department, *Country Tables 1990*, FAO, Rome, 1990. The effect of population pressure in reducing livestock numbers is discussed by Boserup, Ester in *Population and Technology*, Blackwell, Oxford, 1981, pp. 17–18.

25 Irrigated area and livestock number trends calculated from Food and Agriculture Organization, Economic and Social Policy Department, *Country Tables 1990*, FAO, Rome, 1990.

26 The following account is based on Higgins, Graham *et al.*, *Potential Population Supporting Capacities of Lands in the Developing World*, FAO, Rome, 1982; Harrison, Paul, *Land, Food and People*, FAO, Rome, 1984; other material from *Agro-Ecological Zones Project*, vols 1–4, World Soil Resources Report 48, FAO, Rome, 1982.

27 See Harrison *op. cit.*, p. 14, adjusted to fertilizer levels from Food and Agriculture Organization, *Fertilizer Yearbook 1989*, FAO, Rome, 1990.

28 Results with one-third deduction, and for actually cultivated lands, from Harrison, Paul, *op. cit.*, 1984, pp. 34–45.

29 Catch statistics from FAO, *Country Tables 1990*, FAO, Rome, 1990; and FAO *Fishery Statistics: Catches and Landings 1988*, FAO, Rome, 1990.

30 *Ibid.*

31 Whale catches from United Nations Environment Programme, *Environmental Data Report*, UNEP, Blackwell Reference, Oxford, 1989, p. 292.

32 Sustainable yields: Alexandratos, Nikos, ed., *World Agriculture: Toward 2000*, Belhaven Press, London 1988; Robinson, M. A., *Trends and Prospects in World Fisheries*, FAO, Rome, 1984; 1988 catches from FAO, *Yearbook of Fishery Statistics: Catches and Landings 1988*, FAO, Rome, 1990.

33 Food and Agriculture Organization, *Trends and Prospects for Capture Fisheries*, Committee on Fisheries, COFI/89/2, FAO, Rome, 1988.

34 Global water figures from Speidel, David *et al.*, *Perspectives on Water Uses and Abuses*, Oxford University Press, Oxford, 1988, p. 28; current water use from World Resources Institute, *World Resources 1990–91*, World Resources Institute, Washington DC, 1990, p. 330.

35 World Resources Institute, *loc. cit.*

36 Consumption per person for industrial uses calculated from per capita withdrawals and sectoral shares in World Resources Institute, *loc. cit.*

37 Sectoral shares from World Resources Institute, *loc. cit.*; domestic use

by source from World Resources Institute, *World Resources 1986*, World Resources Institute, Washington DC, 1986, p. 130.

38   Falling use in some developed countries: Organization for Economic Cooperation and Development, *Environmental Indicators*, OECD, Paris, 1991, p. 25.

39   Falkenmark, Malin *et al.*, 'Macro-Scale Water Scarcity Requires Micro-Scale Approaches', *Natural Resources Forum*, November 1989, pp. 258–67.

40   Percentage use levels from World Resources Institute, *World Resources 1990–91*, World Resources Institute, Washington DC, 1990, pp. 330–1.

41   *Ibid.*

42   Falkenmark, Malin, 'The Massive Water Scarcity Now Threatening Africa', *Ambio* 18(2), pp. 112–18.

43   These calculations are based on water use figures in World Resources Institute, *World Resources 1990–91*, World Resources Institute, Washington DC, 1990, pp. 330–1. Population projections for 2025 are from the United Nations Population Division, for 2100 from the World Bank. It is assumed that per capita use in agriculture remains stable. Industrial and domestic uses rise to the European levels of 392 and 94 cubic metres per person.

44   Postel, Sandra, *Water for Agriculture*, Worldwatch Paper 93, Worldwatch Institute, Washington DC, 1989.

4   The passing of biological diversity

1   Cowen, Richard, *The History of Life*, Blackwell Scientific Publications, Cambridge, MA, 1990, pp. 97–114. The lowest level of taxonomic classification is the species. Next comes the genus. Then family. Family is a level at which significant formal differences occur, and is often used to measure the impact of extinctions.

2   Stanley, Steven, *Extinctions*, Scientific American Books, New York, 1987.

3   See the overview articles by Walter Alvarez and Vincent Courtillot in *Scientific American*, August 1990, pp. 42–60.

4   The following sections are based on Cowen, *op. cit.*, and Futuyma, Douglas, *Evolutionary Biology*, Sinauer Associates, Sunderland, MA, 1986.

5   For an excellent overview, see Howe, Henry and Westley, Lynn, *Ecological Relationships of Plants and Animals*, Oxford University Press, Oxford, 1988.

6   Stork, Nigel and Gaston, Kevin, 'Counting Species One by One', *New Scientist*, 11 August 1990, pp. 43–7.

7 *op. cit.*

8 Wilson, E. O., ed., *Biodiversity*, National Academy Press, Washington DC, 1988, pp. 4–5.

9 Mammals: Reid, Walter and Miller, Kenton, *Keeping Options Alive*, World Resources Institute, Washington DC, 1989, pp. 10–11. Birds: Ricklefs, Robert, *Ecology*, W. H. Freeman, New York, 1990, p. 750. Oceans: Reid and Miller, *op. cit.* p. 17.

10 Species per 0·1 ha: Mooney, Harold, 'Lessons from Mediterranean Climates Regions', in Wilson, E. O., ed., *Biodiversity*, pp. 9, 157–65.

11 Futuyma, Douglas, *Evolutionary Biology*, Sinauer, Sunderland, MA, 1986, p. 359.

12 Futuyma, *op. cit.*, p. 359, 362, 397; Wilson, *op. cit.*, p. 8.

13 Cowen, Richard, *The History of Life*, Blackwell Scientific Publications, Boston, 1990, pp. 440–54; Stanley, Steven, *Extinction*, Scientific American Books, New York, 1987 pp. 203–7.

14 McNeely, Jeffrey *et al.*, *Conserving the World's Biological Diversity*, IUCN, Gland, Switzerland, 1990, p. 42.

15 Vitousek, P. *et al.*, 'Human Appropriation of the Products of Photosynthesis', *BioScience* 36(6), pp. 368–73.

16 Extinction rates calculated from unpublished data supplied by the World Conservation Monitoring Centre, Cambridge, November 1990.

17 World Conservation Monitoring Centre, unpublished, *op. cit.*

18 Stephen Edwards, International Union for the Conservation of Nature, personal communication.

19 World Conservation Monitoring Centre, unpublished, *op. cit.*

20 Search failures: Reid, Walter and Miller, Kenton, *Keeping Options Alive*, World Resources Institute, Washington DC, 1989, pp. 34–5; McNeely, Jeffrey *et al.*, *Conserving the World's Biological Diversity*, IUCN, Gland, 1990, p. 41.

21 Ehrlich, Paul and Anne, *Extinctions*, Random House, New York, 1981; *Global 2000 Report to the President*, Council on Environmental Quality, Washington DC, 1982; Reid, Walter and Miller, Kenton, *Keeping Options Alive*, World Resources Institute, Washington DC, 1989; Wilson, E. O., ed., *Biodiversity*, National Academy Press, Washington DC, 1988.

22 Total number from World Conservation Monitoring Centre, *op. cit.*, proportion threatened based on McNeely, Jeffrey *et al.*, *Conserving the World's Biological Diversity*, IUCN, Gland, 1990, p. 41, compared with total species numbers from Wilson, E. O., ed., *Biodiversity*, National Academy Press, Washington DC, 1988, pp. 4–5.

23 Reid, Walter and Miller, Kenton, *Keeping Options Alive*, World Resources Institute, Washington DC, 1989, p. 42; Davis, Stephen D.

et al., *Plants in Danger: What Do We Know?*, International Union for the Conservation of Nature, Gland, 1986, p. xxix.

24 Reid and Miller *op. cit.*, p. 40.

25 Tudge, Colin, 'Under Water, Out of Mind', *New Scientist*, 3 November 1990, pp. 40–5.

26 Salm, Rodney, 'Coral Reefs of the Western Indian Ocean', *Ambio* 12(6), pp. 349–53, 1983.

27 Linden, Olof, 'Human Impact on Tropical Coastal Zones', *Nature and Resources* 26(4), pp. 3–11, 1990; World Resources Institute, *World Resources 1986*, World Resources Institute, Washington DC, 1986, p .151.

28 Organization for Economic Cooperation and Development, *Agricultural and Environment Policies*, OECD, Paris, 1989, pp. 36–41, 171.

29 *The African Elephant*, UNEP/GEMS Environment Library no. 3, UNEP, Nairobi, 1989.

30 Proportions of threat due to each cause from Prescott-Allen, Robert, cited in Reid and Miller, *op. cit.*, pp. 41, 45, 50, 51.

31 Organization for Economic Cooperation and Development, *Agricultural and Environment Policies*, OECD, Paris, 1989, pp. 36–41, 171.

32 McNeeley *et al.*, *op. cit.*, p. 47. Figures on habitat loss are from International Union for the Conservation of Nature, *Review of the Protected Areas System in the Afrotropical Realm*, IUCN, Gland, 1986; and IUCN, *Review of the Protected Areas System in the Indo-Malayan Realm*, IUCN, Gland, 1986.

33 The results are as follows for 50 countries grouped in quintiles in terms of percentage habitat loss.

Countries	habitat loss (mid 80s)	persons/km^2 (1989)
Top 20%	85%	189
2nd 20%	78%	119
3rd 20%	67%	45
4th 20%	55%	38
5th 20%	41%	29

(r = 0·486; p < ·001).

Calculated from McNeely, Jeffrey *et al.*, *Conserving the World's Biological Diversity*, IUCN, Gland, 1990, pp. 46–7 (habitat loss), and World Resources Institute, *World Resources 1990–91*, World Resources Institute, Washington DC, 1990, pp. 268–9 (population density).

34 This section is based on personal research in Rwanda and on Weber, William, *Ruhengeri and its Resources*, USAID, Kigali, 1987.

35 The species–area rule states that the number of species rises – or falls – as approximately the fourth root of the increase or decline in area. Actual values range from between the third and the fifth root. For

details, see, Ricklefs, Robert, *Ecology*, 3rd edn, W. H. Freeman, New York, 1990, pp. 723–6.

36 Wilson, ed., *op. cit.*, pp. 12–13.

37 Stork, Nigel and Gaston, Kevin, 'Counting Species One by One', *New Scientist*, 11 August 1990, pp. 43–7; Erwin, Terry, 'The Tropical Forest Canopy', in Wilson, E. O., ed., *Biodiversity*, National Academy Press, Washington DC, 1988, pp. 123–9.

38 Wilson, ed., *op. cit.*, p. 12.

39 International Union for the Conservation of Nature, *The IUCN Plant Red Data Book*, IUCN, Gland, Switzerland, 1978, pp. 13–17.

5 Madagascar

1 Mack, John, *Madagascar*, British Museum, London, 1986.

2 Wildlife: Jolly, Alison *et al.*, eds, *Madagascar*, Pergamon Press, Oxford, 1984; Jenkins, M. D., ed., *Madagascar, an Environmental Profile*, International Union for the Conservation of Nature, Gland, 1987; IUCN, *Priorités en matière de conservation des espèces à Madagascar*, Gland, 1987.

3 Cowen, Richard, *History of Life*, Blackwell Scientific Publications, Boston, MA, 1990, pp. 449–50.

4 Green, Glen and Sussman, Robert, 'Deforestation History of the Eastern Rain Forests of Madagascar from Satellite Images', *Science* 248, pp. 212–5, 1990.

5 Economic data from World Bank, *World Development Report 1990*, World Bank, Washington DC, 1990.

6 Pryor, Frederick, *Malawi and Madagascar*, Oxford University Press, Oxford, 1990.

7 I am indebted to Dr Lon Keitlinger (personal communication) for these survey results, which are averages for Ranomafana villages.

6 Deforestation

1 Ssu-ma Chien, *Shih Chi*, 29, in *Records of the Historian*, trans. Burton Watson, Columbia University Press, New York, 1960.

2 Pritchard, James, *Ancient Near Eastern Texts*, 3rd edn, Princeton University Press, 1969, pp. 82–6; supplemented by Sandars, N. K., *The Epic of Gilgamesh*, Penguin, 1972, pp. 70–84.

3   Mather, A. S., *Global Forest Resources*, Belhaven Press, London, 1990.
4   White, Lynn, *Medieval Technology and Social Change*, Oxford University Press, Oxford, 1962.
5   Cox, T. R., *et al.*, *This Well-wooded Land*, University of Nebraska Press, Lincoln and London, 1985; Mather, A. S., *Global Forest Resources*, Belhaven Press, London, 1990.
6   Matthews, Elaine, *Journal of Climate and Applied Meteorology* 22, pp. 475–87, 1983. Matthews's data is based on atlases that date mostly from the 1960s and early 1970s. Houghton R. A., *Ecological Monographs* 53(3), pp. 235–62, 1983 (cited with additional figures in World Resources Institute, *World Resources 1987*, World Resources Institute, Washington DC, 1987, p. 272).
7   Range of estimates from: Mather, A. S., *Global Forest Resources*, Belhaven Press, London, 1990, pp. 59–66; Food and Agriculture Organization, *Production Yearbook 1989*, FAO, Rome, 1990. Almost all forestry statistics from developing countries are unreliable. The FAO statistics are based on returns from governments, and they indicate land that is officially classified as forest, farmland, and so on. For all its deficiencies, this is the only set of data that allows different types of land use to be compared over time. Like all environmental data, it gives a general guide to trends rather than an accurate picture at any one time.
8   Food and Agriculture Organization, *Production Yearbook 1989*, FAO, Rome, 1990.
9   Lanly, Jean-Paul, *Tropical Forest Resources*, FAO Forestry Paper 30, FAO, Rome, 1982.
10 Brazil range and high India figure: World Resources Institute, *World Resources 1990–91*, World Resources Institute, Washington DC, 1990, p. 102; low India figure from: Forest Survey of India, *The State of Forest Report 1989*, Dehra Dun, 1989.
11 Myers, Norman, *Conversion of Tropical Moist Forests*, National Academy of Sciences, Washington DC, 1980; Lanly, Jean-Paul, *Tropical Forest Resources*, FAO Forestry Paper 30, FAO, Rome, 1982. In 1989 Myers produced an updated, much lower figure of 142,000 square kilometres a year, in Myers, Norman, *Deforestation Rates in Tropical Forests and their Climatic Implications*, Friends of the Earth, London, 1989.
12 Food and Agriculture Organization, *Production Yearbook 1989*, FAO, Rome, 1990.
13 Madagascar: Green, Glen and Sussman, Robert, 'Deforestation History of the Eastern Rainforests of Madagascar', *Science* 248, pp. 212–15, 1990; Ivory Coast: Repetto, Robert *et al.*, *The Forest for the Trees*, World Resources Institute, Washington DC, 1988; Philippines:

World Wide Fund for Nature, *Asian Tropical Forests*, WWF, Gland, undated; Ethiopia: Harrison, Paul, *The Greening of Africa*, Penguin and Paladin, New York and London, 1987.

14 Lanly, Jean-Paul, *Tropical Forest Resources*, FAO Forestry Paper 30, FAO, Rome, 1982. According to FAO land-use figures (FAO, *Production Yearbook 1989*, FAO, Rome, 1990) some 86,000 square kilometres of forest were being converted to other uses each year in 1973–8. By 1983–8 the annual rate of clearance was up by a quarter, to 107,000 square kilometres. I have not quoted these figures in the main text to avoid confusion with the FAO's two tropical forest assessments.

15 Country data in this and the three following paragraphs from *Forest Resources 1980*, FAO, Rome, 1985.

16 FAO Forest Resources Assessment Project, *Second Interim Report on the State of Tropical Forests*, 10th World Forestry Congress, Paris, September 1991.

18 Lanly, Jean-Paul, *Tropical Forest Resources*, Forestry Paper 30, FAO, Rome, 1982.

19 Global management figure: World Resources Institute, *World Resources 1986*, World Resources Institute, Washington DC, 1986; tropical figure: Lanly, Jean-Paul, *Tropical Forest Resources*, FAO Forestry Paper 30, FAO, Rome, 1982; Poore, Duncan, ed., *No Timber Without Trees*, Earthscan, London, 1989, p. 196.

20 Whitmore, T. C., *Tropical Rainforests of the Far East*, 2nd edn, Clarendon Press, Oxford, 1984.

21 Mather, A. S., *Global Forest Resources*, Belhaven Press, 1990, pp. 223–8.

22 *op. cit.*, p. 233.

23 Whitmore, T. C., *op. cit.*, note 20, 1984, pp. 272–7.

24 All figures on land-use changes in the following paragraphs are calculated from Food and Agriculture Organization, *Production Yearbook 1989*, FAO, Rome, 1990.

25 Mahar, Denis, *Government Policies and Deforestation in Brazil's Amazon Region*, World Bank, Washington DC, 1989; Repetto, Robert *et al.*, *The Forest for the Trees*, World Resources Institute, Washington DC, 1988.

26 Herd numbers from Food and Agriculture Organization, *Country Tables 1990*, FAO, Rome, 1990; exports and imports from FAO, *Trade Yearbook 1989*, FAO, Rome, 1990.

27 Food and Agriculture Organization, *Forest Products Yearbook 1988*, FAO, Rome, 1990.

28 Mather, A. S., *Global Forest Resources*, Belhaven Press, London 1990; Whitmore, T. C., *Tropical Rainforests of the Far East*, Clarendon Press, Oxford, 1985.

29 Repetto, Robert *et al.*, *The Forest for the Trees*, World Resources Institute, Washington DC, 1988, pp. 17–21.

7 Forest adjustments

1 Food and Agriculture Organization, *Production Yearbook 1989*, FAO, Rome, 1990.

2 Lanly, J. P., *Unasylva* 31(123), pp. 12–20.

3 Whitmore, T. C., *Tropical Rainforests of the Far East*, Clarendon Press, Oxford, 1985, pp. 279–82.

4 Mahar, Denis, *Government Policies and Deforestation in Brazil's Amazon Region*, World Bank, Washington DC, 1989.

5 Bilsborrow, Richard and Geores, Martha, *Population, Environment and Sustainable Agricultural Development*, FAO/Netherlands Conference on Agriculture and the Environment, Rome, 8–12 October 1990, pp. 53–5.

6 Boserup, Ester, *The Conditions of Agricultural Growth*, Allen & Unwin, London, 1965, p. 118.

7 Boserup, *op. cit.*, pp. 41, 118.

8 Duby, Georges, 'Medieval Agriculture, 900–1500', in Cipolla, Carlo, ed., *Fontana Economic History of Europe: The Middle Ages*, Fontana Books, 1972, pp. 175–220.

9 See Harrison, Paul, *The Greening of Africa*, Paladin and Penguin Books, London and New York 1987, *passim*; Lele, Uma and Stone, Steven, *Population Pressure, the Environment and Agricultural Intensification*, World Bank, Washington DC, 1989, examines why Africa has so far failed to follow the Boserup process.

10 Calculated from FAO data files on cereal production, area and yield, Nikos Alexandratos, personal communication.

11 Higgins, G. M. *et al.*, *Potential Population Supporting Capacities of Lands in the Developing World*, FAO, Rome, 1982, pp. 17–19.

12 India: Centre for Science and Environment, 'The Environmental Problems Associated with India's Major Cities', *Environment and Urbanization* 1(1), pp. 7–15, 1989; Mexico City: Schteingart, Martha, 'The Environmental Problems Associated with Urban Development in Mexico City', *Environment and Urbanization* 1(1), pp. 40–50, 1989; São Paulo: Brown, Lester and Jacobson, Jodi, *The Future of Urbanization*, Worldwatch Paper 77, May 1987, p. 19; Egypt and China: FAO Economic and Social Policy Department, *1990 Country Tables* and FAO *Production Yearbook 1989*, both FAO, Rome, 1990.

13 This assumes a projected stable world population of 11·5 billion, as in the UN's medium projection.

14 A correlation of 1 between two factors means that as one of them

increases or decreases, so does the other. A correlation of −1 means that as one increases, the other decreases. Alexander Mather found a correlation (r²) of −0·683 between population growth and forest cover for the 1975–81 period ('Global Trends in Forest Resources', *Geography* 72(1), pp. 1–15, 1987). M. Palo and G. Mery found close correlations between population density and forest cover: −0·8 for Latin America and moist Africa, and −0·96 for Asia (*18th IUFRO Congress Report*, Ljubljana, 1986, pp. 552–85). The FAO's 1990 forest assessment study looked at areas with two measurements of forest cover, and found a close correlation between population density and deforestation – ranging up to 0·96. The relationship was so close that for many countries where only one measurement of forest cover was available population figures were used to assess forest cover at a later date (K. D. Singh *et al.*, *A Model Approach to Studies of Deforestation*, Forestry Department, FAO, 1990).

15 Green, Glen and Sussman, Robert, 'Deforestation History of the Eastern Rain Forests of Madagascar from Satellite Images', *Science* 248, pp. 212–15, 1990.

16 The following land-use figures are taken from Food and Agriculture Organization, *Production Yearbook 1989*, FAO, Rome, 1990.

17 The share of wetlands in the earth's land area is 6 per cent (Maltby, Edward, *Waterlogged Wealth*, Earthscan, 1989). I have assumed that wetlands made up a similar proportion in increased farmland. I have added the 58,000 square kilometres of African pastureland that probably became arable. The consumption share is derived from calculations in the appendix (pp. 306–314).

18 I say rough advisedly. Basic data on area are not reliable for many countries.

19 United Nations Population Division, *Prospects of World Urbanization 1988*, United Nations, New York, 1989.

20 Future growth rates of rural populations from United Nations Population Division, *op. cit.*

21 de Montalembert, M. R., and Clément, J., *Fuelwood Supplies in the Developing Countries*, FAO, Rome, 1983, p. 22.

22 See Harrison, Paul, *The Greening of Africa*, Penguin and Paladin Books, New York and London, 1987, chs 11 and 12

23 Barnard, Geoffrey and Kristoferson, Lars, *Agricultural Residues as Fuel in the Third World*, Earthscan, International Institute for Environment and Development, London, 1985.

24 Grainger, Alan, 'Population as Concept and Parameter in the Modelling of Tropical Land Use Change', in Ness, Gayl, ed., *Proceedings of the International Symposium on Population–Environment Dynamics*, Michigan University Press, 1991.

25 The following section is based on Boserup's basic principle, extended into the field of forestry, with elements from Alexander Mather's model of forest resource and management trends (Mather, A. S., *Global Forest Resources*, Belhaven Press, London, 1990, pp. 31, 90).

8 Land degradation

1 Ssu-Ma Chien, *Shi Chi*, 29, in *Records of the Historian*, trans. Burton Watson, Columbia University Press, New York, 1969.

2 Plato, *Critias*, 111b–d, transl. A. E. Taylor, in Plato, *Collected Dialogues*, Princeton University Press, Princeton, 1963.

3 See Edward Hyams's classic *Soil and Civilization*, Thames and Hudson, London, 1952.

4 United Nations Environment Programme, *General Assessment of Progress in the Implementation of the Plan of Action to Combat Desertification 1978–85*, UNEP, Nairobi, 1984; share calculated from Food and Agriculture Organization, *Production Yearbook 1989*, FAO, Rome, 1990.

5 Africa: Boyagdiev, Todor, *Map of Desertification Hazards*, Explanatory Note, UNEP, May 1984; Near East: Hauck, F. W., 'Soil Erosion and its Control in Developing Countries', in El-Swaify, S. A., ed., *Soil Erosion and Conservation*, Soil Conservation Society of America, Iowa, 1985, pp. 718–28.

6 L. R. Oldeman *et al.*, *World Map of the Status of Human-Induced Soil Degradation*, Global Assessment of Soil Degradation, International Soil Reference and Information Centre, Wageningen, Netherlands, 1990.

7 Quoted in United States Department of Agriculture, Economic Research Service, *World Agriculture Situation and Outlook Report*, Washington DC, 1989.

8 Nelson, Ridley, *Dryland Management: The Desertification Problem*, World Bank, Washington DC, 1988.

9 Mabbutt, J. A., A new Global Assessment of the Status and Trends of Desertification, *Environmental Conservation* 11(2), pp. 103–13.

10 The 1990 population figure assumes the average rural population growth rate of 1·46 per cent between 1985 and 1990.

11 'Moderate' desertification was defined as a loss of 'up to' 25 per cent loss of productivity. This could include land where the loss was as low as 1 per cent. Africa was undergoing a severe drought at the time of the assessment. Many countries did not carry out detailed studies, so expert guesstimates were made. See Nelson, Ridley, *Dryland Management: The Desertification Problem*, World Bank, Washington DC, 1988.

12 See Harrison, Paul, *The Greening of Africa*, Penguin and Paladin, New York and London, 1987, pp. 142–6.

13 Kishk, Mohammed Atif, Desert Encroachment in Egypt's Nile Valley, in El-Swaify, S. A., ed., *Soil Erosion and Conservation*, Soil Conservation Society of America, Iowa, 1985.

14 World Resources Institute, *World Resources 1986*, World Resources Institute, Washington DC, 1986, p. 81; World Resources Institute, *World Resources 1990–91*, World Resources Institute, Washington DC, 1990, p. 114.

15 *Ibid.*

16 Nelson, Ridley, *Dryland Management: The Desertification Problem*, World Bank, Washington DC, 1988, p. 10.

17 United States: Organization for Economic Cooperation and Development, *The State of the Environment*, OECD, Paris, 1991, p. 97; India: El-Swaify, *op. cit.*, pp. 3–8; other examples: World Resources Institute, *World Resources 1988–89*, Washington DC, 1988, p. 282.

18 Impact on US yields: Crosson, Pierre, 'Impact of Erosion on Land Productivity', in El-Swaify, *op. cit.*, pp. 217–36; Nigeria: Lal, Rattan, 'Soil Erosion and Productivity in Tropical Soils', in El-Swaify, *op. cit.*, p. 244.

19 Lal, Rattan, *No-Till Farming*, Monograph no. 2, International Institute of Tropical Agriculture, Ibadan, 1983.

20 Harrison, Paul, *Land, Food and People*, FAO, Rome, 1984, pp. 9–10.

21 US costs: Crosson, Pierre, 'Impact of Erosion on Land Productivity', in El-Swaify, *op. cit.*, pp. 217–36; sediment figures: Meybeck, Michel *et al.*, *Global Freshwater Quality*, Blackwell Reference, Oxford, 1990, p. 104; off-farm costs: Conservation Foundation, reported in Organization for Economic Cooperation and Development, *Agricultural and Environmental Policies*, OECD, Paris, 1989.

22 Food and Agriculture Organization, *An International Action Programme on Water and Sustainable Development*, FAO, Rome, 1990, p. 21.

23 Meybeck, Michel *et al.*, *Global Freshwater Quality*, Blackwell Reference, Oxford, 1990, pp. 149–56.

24 Tandon, H. L. S., 'Where Rice Devours the Land', *Ceres* 126, November 1990, pp. 25–9.

25 Smaling, Eric, 'Two Scenarios for the Sub-Sahara', *Ceres* 126, November 1990, pp. 19–24.

26 Tandon, H. L. S., *op. cit.*

27 Blaikie, Piers and Brookfield, Harold, *Land Degradation and Society*, Methuen, London, 1987, p. 85.

28 Simon, Julian, *The Ultimate Resource*, 1980 p. 81; *op. cit.*, 1984, pp. 21–3. Food and Agriculture Organization, *Production Yearbook 1989*, FAO, Rome.

29 Lappé, Frances Moore and Collins, Joseph, *Food First*, Souvenir Press, London, 1977, pp. 40–9.

30 Blaikie, Piers, *The Political Economy of Soil Erosion*, Longman, London, 1985; Blaikie, Piers and Brookfield, Harold, *Land Degradation and Society*, Methuen, London, 1987.

31 All figures calculated by the author from FAO, *Production Yearbooks, 1981* and *1985*, in background paper for World Bank, *Sub-Saharan Africa: From Crisis to Sustainable Growth*, World Bank, Washington DC, 1989.

32 *National Agricultural Survey*, Ministry of Agriculture, Kigali, 1986.

9 Living on the margin

1 Columella, *De Re Rustica*, Preface, 2–3.

2 Sinha, Rada, *Landlessness: A Growing Problem*, FAO, 1984, pp. 16–17.

3 See pp. 221–35. Higgins, Graham *et al.*, *Potential Population Supporting Capacities of Lands in the Developing World*, FAO, Rome, 1982; Harrison, Paul, *Land, Food and People*, FAO, Rome, 1984. The FAO analysis was made not only for countries, but for small zones *within* countries that had the same climate and soil characteristics. The critical zones are climatic zones, cutting across country boundaries, that could not support their 1975 populations.

4 World Bank, *World Development Report 1990*, World Bank, Washington DC, 1990, p. 29; Leonard, H. Jeffrey *et al.*, *Environment and the Poor*, Transaction Books, New Brunswick, 1989.

5 The arguments and evidence on both sides are admirably marshalled in Singh, Inderjit, *The Great Ascent*, Johns Hopkins, Baltimore, 1990.

6 *Statistical Outline of India 1989–90*, Tata Services Ltd, Bombay, 1990.

7 Attwood, D. W., 'Why Some of the Poor Get Richer', *Current Anthropology* 20(2), pp. 495–515, 1979.

8 Singh, Inderjit, *The Great Ascent*, Johns Hopkins, Baltimore, 1990, p. 60.

9 Shift in landholding patterns from: United Nations Development Programme, *Bangladesh Agricultural Sector Review*, UNDP, Dhaka, 1989.

10 Bangladesh Bureau of Statistics, *Statistical Pocketbook of Bangladesh 1990*, Dhaka, 1990, p. 125.

11 Effects of one generation's growth assumes 1988 rice yields of 2·4 tonnes per hectare, and 1990 total fertility rate of five children per woman. Data from Bangladesh Bureau of Statistics, *Statistical Pocketbook of Bangladesh 1990*, Dhaka, 1990, p. 125.

12 Parthasarathy, G., *Understanding Agriculture: Growth, Structure and Challenges*, mimeo, FAO, Rome, March 1979.

13 Wilson, Francis and Ramphele, Mamphela, *Uprooting Poverty In South Africa*, Norton, New York, 1989.

14   Alvares, Claude and Billorey, Ramesh, *Damming the Narmada*, Third World Network, Penang, 1988.
15   Lele, Uma and Stone, Steven, *Population Pressure, the Environment*, and Agricultural Intensification, World Bank, Washington DC, 1989, p. 22.
16   Talbot, Lee, 'Demographic Factors in Resource Depletion and Environmental Degradation in an East African Rangeland', *Population and Development Review* 12(3), pp. 441–51, 1986.
17   This and other data from: Conservation Division, *National Conservation Plan for Lesotho*, Ministry of Agriculture, Maseru, 1988.
18   This section is based on research for my report to King Moshoeshoe II of Lesotho, *The Greening of Lesotho*, Oak Foundation, Gland, 1989.
19   Hudson, Norman, *Soil Conservation*, Batsford, London, 1985, p. 198.
20   Increase in cropped area leads to longer slopes, assuming that no terraces are built – that is, holding *technology* constant.
21   Livestock figures from *FAO Production Yearbook 1979* and *1989*, FAO.
22   FAO, *Atlas of African Agriculture*, FAO, Rome, 1986, p. 30.
23   Barnard, Geoffrey and Kristoferson, Lars, *Agricultural Residues as Fuel in the Third World*, Earthscan, International Institute for Environment and Development, London 1985, pp. 29–30, 41; World Bank, Bangladesh: Rural and Renewable Energy Development Prospects, Energy Department Paper no. 5, World Bank, Washington DC, 1982.

10   Burkina Faso

1   For more detail see Harrison, Paul, *The Greening of Africa*, Penguin and Paladin, New York and London, 1987.

12   The environmental impact of cities

1   Varro, *De Re Rustica*, ii.3.
2   Athens: Zimmern, Alfred, *The Greek Commonwealth*, 5th edn, Oxford University Press, Oxford, 1931, pp. 174–9; Rome: Cowell, F. R., *Cicero and the Roman Republic*, Penguin, 1973, pp. 61–2.
3   Juvenal, *Satires*, iii *passim*, especially lines 232–48, 254–61 and 268–314; Martial, *Epigrams*, i.86.
4   Asian cities: Chandler, Tertius and Fox, Gerald, *3000 Years of Urban Growth*, Academic Press, New York, 1974; Constantinople: Cipolla, Carlo, ed., *The Middle Ages*, Fontana Economic History of Europe, London, 1974, p. 34; cities in 1500: Cipolla, Carlo, ed., *The Sixteenth*

and Seventeenth Centuries, Fontana Economic History of Europe, London, 1974, pp. 42–3.

5   The United Nations Population Division accepts national assessments of the proportion living in towns, but the definition of urban varies widely between countries. In Sweden 200 or more people constitute an urban settlement. In Nigeria a community must number over 20,000 before it qualifies for the name urban. For the present we have little choice but to use United Nations data, bearing in mind that they are most useful to indicate trends over time rather than to compare urbanization levels between countries. See Hardoy, Jorge and Satterthwaite, David, 'Urban Change in the Third World', *Habitat International* 10(3), pp. 33–52, 1986.

6   Figures for urban growth are from United Nations Population Division, *World Urbanization Prospects 1990*, United Nations, New York, 1991.

7   United Nations Population Division, *World Urbanization Prospects 1990*, ST/ESA/SER A/112, United Nations, New York, 1991.

8   Bairoch, Paul, *Urban Unemployment in Developing Countries*, ILO, Geneva, 1973, p. 19; United Nations Population Division, *Prospects of World Urbanization 1988*, ST/ESA/SER.A/112, United Nations, New York, 1989.

9   Calculated from World Bank, *World Development Report 1990*, World Bank, Washington DC, 1990, pp. 238–9.

10   Hardoy, Jorge and Satterthwaite, David, *Small and Intermediate Urban Centres*, Hodder & Stoughton, London, 1986, pp. 326–7.

11   *Ibid.*

12   United Nations Population Division, *World Urbanization Prospects 1990*, United Nations, New York, 1991, forthcoming.

13   Regional averages calculated from Grimes, Orville, *Housing for Low Income Families*, Johns Hopkins University Press, 1976, pp. 116–17.

14   Individual city figures: United Nations Centre for Human Settlements, *Global Report on Human Settlements*, Oxford University Press, Oxford, 1987, p. 77.

15   See Harrison, Paul, *The Third World Tomorrow*, Penguin Books, 1980, pp. 104–9.

16   United Nations Centre for Human Settlements, *Global Report on Human Settlements 1986*, Oxford University Press, Oxford, 1987, p. 77, and annexe, Table 14.

17   United Nations Population Division, *World Population Trends and Policies: 1989 Monitoring Report*, ESA/P/WP.107, United Nations, New York, 1989, Table 80; Kenya figures from Bubba, Ndinda and Lamba, Davinder, 'Local Government in Kenya', *Environment and Urbanization* 3(1), pp. 37–59, 1991.

18 Schteingart, Martha, 'The Environmental Problems Associated with Urban Development in Mexico City', *Environment and Urbanization* 1(1), pp. 40–50, 1989; United Nations Centre for Human Settlements, *Global Report on Human Settlements 1986*, Oxford University Press, Oxford, 1987, p. 83.

19 For open unemployment levels see Kahnert, Friedrich, *Improving Urban Employment and Labor Productivity*, World Bank Discussion Paper 10, World Bank, Washington DC, 1987; and International Labour Office, *Bulletin of Labour Statistics 1991–1*, International Labour Office, Geneva, 1991.

20 The role of the informal sector in modifying Todaro's thesis is discussed in Mazumdar, D., *The Theory of Urban Unemployment in Less Developed Countries*, World Bank Staff Working Paper no. 198, Washington DC, 1975; see also Turnham, David, ed., *The Informal Sector Revisited*, Development Centre, OECD, Paris, 1990, pp. 10–48; the regional extent of the informal sector is taken from p. 22 (Latin America) and p. 19 (averages for sub-Saharan Africa and Asia).

21 Turnham, *op. cit.*, pp. 36–9.

22 WHO Community Water Supply Unit, *The International Drinking Water Supply and Sanitation Decade: Review of Decade Progress as at December 1988*, Division of Environmental Health, World Health Organization, Geneva, 1990, p. 19.

23 Regional sanitation figures from World Health Organization, *op. cit.*, p. 19; India: Centre for Science and Environment, 'The Environmental Problems Associated with India's Major Cities', *Environment and Urbanization* 1(1), pp. 7–15, 1989.

24 Mexico: Schteingart, Martha, 'The Environmental Problems associated with urban development in Mexico City', *Environment and Urbanization* 1(1), pp. 40–50, 1989.

25 Mitchell, R. J., and Leys, M. D. R., *A History of London Life*, Penguin, London, 1963, pp. 171–2.

26 Global Environment Monitoring System, *Assessment of Urban Air Quality*, United Nations Environment Programme, Nairobi, 1988, p. 15.

27 United Nations Environment Programme, *Environmental Data Report*, 2nd edn, Blackwell Reference, Oxford, 1989, p. 15 and Table 1.17.

28 Hardoy and Satterthwaite, *Squatter Citizen*, *op. cit.*, p. 196.

29 Vendor prices from: World Bank Urban Development Division, *Urban Strategy Paper*, World Bank, Washington DC, 1989, p. 70.

30 Basta, Samir, 'Nutrition and Health in Low Income Urban Areas of the Third World', *Ecology of Food and Nutrition*, 6, pp. 113–24.

31 United Nations Centre for Human Settlements, *Global Report on Human Settlements 1986*, Oxford University Press, Oxford, 1987, p. 83.

32 Travel costs from United Nations Centre for Human Settlements, *Global Report on Human Settlements 1986*, Oxford University Press, Oxford, 1987, p. 83; car figures from Linn, Johannes, *Cities in the Developing World*, Oxford University Press, Oxford, 1983, pp. 95–8.

33 Basta, Samir, 'Nutrition and Health in Low Income Urban Areas of the Third World', *Ecology of Food and Nutrition*, 6, pp. 113–24.

34 A good overall review is Brown, Lester and Jacobson, Jodi, *The Future of Urbanization*, Worldwatch Paper 77, Worldwatch Institute, Washington DC 1987.

35 Food and Agriculture Organization, *State of Food and Agriculture 1985*, FAO, Rome, 1986, pp. 69–75.

36 Bowonder, B. *et al.*, *Deforestation and Fuelwood Use in Urban Centres*, Centre for Energy, Environment and Technology, Hyderabad, 1985; Centre for Science and Environment, *The State of India's Environment 1984–5*, New Delhi, 1982, pp. 20, 28.

37 United Nations Population Division, *Prospects of World Urbanization 1990*, ST/ESA/SER A/112, United Nations, New York, 1991. For 97 developing countries where data are available, the urban population growth rate for 1975–80 correlated with national population growth rate (r = ·673, p < ·001). The rate of urbanization – that is, the annual rate of increase in the urban share of total population – also correlated with the rate of national population growth, though less strongly (r = 0·295, p < ·01). Most of the anomalies are explained by differences in the pre-existing level of urbanization.

38 Figures for 1960s and 1970s are from United Nations Population Division, *Patterns of Urban and Rural Population Growth*, New York, 1980. For 1980–5 the contribution of migration is calculated from United Nations Population Division, *Prospects of World Urbanization 1988*, ST/ESA/SER A/112, United Nations, New York, 1989, using the formula:

Migration share = (Urban growth rate − national growth rate) × 100 ÷ urban growth rate

This first approximation method does not allow for differences in fertility and mortality between urban and rural areas, or for reclassification of urban centres as they pass a certain threshold. It probably overstates the share of migration somewhat. See Renaud, Bertrand, *National Urbanization Policy In Developing Countries*, Oxford University Press, Oxford, 1981, pp. 29–31, 165–9.

39 World Bank, *Education,* Sector Working Paper, World Bank, Washington DC, 1974.

40 Linn, Johannes, *Cities in the Developing World*, Oxford University Press, Oxford, 1983, pp. 23–5.

41 WHO Community Water Supply Unit, *The International Drinking Water Supply and Sanitation Decade: Review of Decade Progress as at December 1988*, Division of Environmental Health, World Health Organization, Geneva, 1990, p. 19.

42 Infant mortality comparisons: World Bank, *World Development Report 1990*, World Bank, Washington DC, 1990, p. 31; life expectancy in Bangladesh: Bangladesh Bureau of Statistics, *Statistical Pocket Book of Bangladesh 1990*, p. 90.

43 Bairoch data calculated from Bairoch, Paul, *Urban Unemployment in Developing Countries*, ILO, Geneva, 1973, p. 29; later data from Lecaillon, Jacques *et al.*, *Income Distribution and Economic Development*, International Labour Office, Geneva, 1984, pp. 54–6.

44 GDP growth rates from: World Bank, *World Development Report 1990*, World Bank, Washington DC, 1991, pp. 180–1.

45 Personal communication, Ministry of Agriculture, Lesotho.

46 Kahnert, Friedrich, *Improving Urban Employment and Labor Productivity*, World Bank Discussion Paper 10, World Bank, Washington DC, 1987, p. 61.

47 See Hardoy and Satterthwaite, *Squatter Citizen*, *op. cit.*

48 Green Revolution labour requirements: see Mahabub Hossain, *The Nature and Impact of the Green Revolution in Bangladesh*, Research Report 67, International Food Policy Research Institute, Washington DC, 1988. Paraná, see Mahar, Dennis, *Government Policies and Deforestation in Brazil's Amazon Region*, World Bank, Washington DC, 1989.

49 United Nations Population Division, *Prospects of World Urbanization 1988*, ST/ESA/SER.A/112, United Nations, New York, 1989.

50 David Satterthwaite, personal communication.

51 Mitchell, R. J., and Leys, M. D. R., *A History of London Life*, Penguin, London, 1963, pp. 267–74.

13 Waste

1 From Benedict, Ruth, *Patterns of Culture*, Routledge & Kegan Paul, London, 1935, p. 143.

2 Mitchell, R. J., and Leys, M. D. R., *A History of London Life*, Penguin, London, 1963, pp. 53–5.

3 Young, John E., *Discarding the Throwaway Society*, Worldwatch Paper 101, Worldwatch Institute, Washington DC, 1991, pp. 22–3; Pollock, Cynthia, *Mining Urban Wastes*, Worldwatch Paper 76, Worldwatch Institute, Washington DC, 1987, pp. 8–12.

4 The typical figures given for low- and middle-income countries are the median of the range of 0·4–0·6 and 0·5–0·9 kilos per day quoted in

Cointreau, Sandra, *Integrated Resource Recovery*, World Bank Technical Paper no. 10, World Bank, Washington DC, 1984, p. 2. Western country figures are from: Organization for Economic Cooperation and Development, *Environmental Indicators*, OECD, Paris, 1991 p. 45, and OECD, *The State of the Environment*, OECD, Paris, 1991, pp. 146–9.

5 Port Harcourt: Izeogu, C. V., 'Urban Development and the Environment in Port Harcourt', *Environment and Urbanization* 1(1), pp. 64, 1989; Cairo: Cointreau, Sandra, *Recycling from Municipal Refuse*, World Bank technical paper 30, World Bank, Washington DC, 1984, p. 6.

6 Cointreau, Sandra, *Integrated Resource Recovery*, World Bank Technical Paper no. 10, World Bank, Washington DC, 1984, p. 2.

7 Cointreau, Sandra, *Environmental Management of Urban Solid Waste in Developing Countries*, World Bank, Washington DC, 1982, p. 14.

8 Figures calculated from Organization for Economic Cooperation and Development *The State of the Environment*, OECD, Paris, 1991, p. 146. They may not always be strictly comparable – for example the USA figure includes liquid wastes managed in land-based operations.

9 *Ibid.*

10 *Op. cit.* p. 145.

11 *Op. cit.* p. 149. The population contribution is calculated by deducting the rise in per capita output of municipal waste from the total output. See appendix for details of this method.

12 These figures are based on the following data:

MUNICIPAL WASTE PRODUCTION

	Developing countries		Europe	N. America
	Low income	Middle income		
per year	182 kg	255 kg	336 kg	826 kg
life expectancy	60 yrs	66 yrs	76 yrs	76 yrs
per life	10·9 tn	16·8 tn	25·5 tn	62·8 tn
% moist	60	50	30	30
dry production	4·36 tn	8·4 tn	17·85 tn	43·96 tn

INDUSTRIAL WASTE

	Developing countries	Europe	N. America
per year	0·046 tn	0·66 tn	3·02 tn
per life	2·9 tn	50·16 tn	229·5 tn

COMBINED WEIGHT

		Europe	N. America
per year	264·5 kg	996 kg	3846 kg
per lifetime	16·66 tn	75·66 tn	292·3 tn

MULTIPLE OF BODYWEIGHT

149x   971x   3907x

The size of the junk cube is based on average refuse densities of 150 kilos per cubic metre.

Figures based on: Cointreau, Sandra, *Integrated Resource Recovery*, World Bank Technical Paper no. 10, World Bank, Washington DC, 1984, p. 2 (low- and mid-income countries); for Europe and North America Organization for Economic Cooperation and Development, *Environmental Indicators*, OECD, Paris, 1991 p. 45; Organization for Economic Cooperation and Development *The State of the Environment*, OECD, Paris, 1991, pp. 149.

13   Phantumvanit, Dhira *et al.*, 'Coming To Terms with Bangkok's Environmental Problems', *Environment and Urbanization* 1(1), pp. 31–39, 1989; Jimenez, Rosario, Metropolitan Manila, *Environment and Urbanization* 1(1), pp. 51–8, 1989.

14   Mayhew, Henry, *London Labour and the London Poor*, 4 vols, 1851–61.

15   Cointreau, Sandra, *Solid Waste Recycling: Case Studies in Developing Countries*, World Bank, Washington DC, 1987, pp. 39, 66.

16   Landfill details: United Nations Environment Programme, *Environmental Data Report 1989/90*, p. 455. Dirty landfill sites: Organization for Economic Cooperation and Development *The State of the Environment*, OECD, Paris, 1991, pp. 152.

17   Organization for Economic Cooperation and Development, *op. cit.*, p. 152.

18   Incineration levels from: United Nations Environment Programme, *Environmental Data Report 1989/90*, p. 455; Organization for Economic Cooperation and Development *The State of the Environment*, OECD, Paris, 1991, p. 154; Young, John E., *Discarding the Throwaway Society*, Worldwatch Paper 101, Worldwatch Institute, Washington DC, 1991, p. 18.

19   Savings from: Cointreau, Sandra, *Recycling from Municipal Waste*, World Bank Technical Paper 30, World Bank, Washington DC, 1984, pp. 3–4.

20   This and following recycling data and trends from: United Nations Environment Programme, *Environmental Data Report 1989/90*, pp. 470–8. OECD Figures from Organization for Economic Cooperation and Development *The State of the Environment*, OECD, Paris, 1991, pp. 153.

21   Young, John E., *Discarding the Throwaway Society*, Worldwatch Paper 101, Worldwatch Institute, Washington DC, 1991, pp. 23, 25; Pollock, Cynthia, *Mining Urban Wastes*, Worldwatch Paper 76, Worldwatch Institute, Washington DC, 1987, pp. 8–12.

22 Organization for Economic Cooperation and Development *The State of the Environment*, OECD, Paris, 1991, pp. 153 and 149.

14 Polluted waters

1 Meybeck, Michael *et al.*, *Global Freshwater Quality*, United Nations Environment Programme, Blackwell Reference, Oxford, 1990, p. 288.

2 Brown, Lester and Jacobson, Jodi, *The Future of Urbanization*, Worldwatch Paper 77, Worldwatch Institute, Washington DC, 1987.

3 Organization for Economic Cooperation and Development, *Agricultural and Environmental Policies*, OECD, Paris, 1989, pp. 42–7 and 146–59.

4 OECD figures from: Organization for Economic Cooperation and Development, *The State of the Environment*, OCED, Paris, 1991, p. 58; rest from: World Health Organization, *The International Drinking Water Supply and Sanitation Decade, Review of Decade Progress*, WHO, Geneva, 1990, p. 19.

5 United Nations Economic Commission for Latin America, *The Water Resources of Latin America and the Caribbean: Water Pollution*, ECLAC, Santiago, Chile, 1989, p. 5; Centre for Science and Environment, 'The Environmental Problems Associated with India's Major Cities', *Environment and Urbanization* 1(1), pp. 7–15, 1989; China: Meybeck, Michael, *et al.*, *Global Freshwater Quality*, United Nations Environment Programme, Blackwell Reference, Oxford, 1990, p. 91.

6 Organization for Economic Cooperation and Development, *The State of the Environment*, OECD, Paris, 1991, pp. 57–9; United Nations Environment Programme, *Environmental Data Report*, UNEP, Blackwell Reference, Oxford, 1989, p. 457.

7 United Nations Environment Programme, *The State of the Marine Environment*, UNEP Regional Seas Reports and Studies no. 115, UNEP, Nairobi, 1990, p. 14.

8 Meybeck, Michael, *et al.*, *Global Freshwater Quality*, United Nations Environment Programme, Blackwell Reference, Oxford, 1990, pp. 46–7.

9 United Nations Environment Programme, *The State of the Marine Environment*, UNEP Regional Seas Reports and Studies no. 115, UNEP, Nairobi, 1990, pp. 17 and 20.

10 Organization for Economic Cooperation and Development, *The State of the Environment*, OCED, Paris, 1991, p. 181.

11 Calculated from Food and Agriculture Organization, *Fertilizer Yearbook 1989*, FAO, Rome, 1990.

12 Meybeck, Michael *et al.*, *Global Freshwater Quality*, United Nations

Environment Programme, Blackwell Reference, Oxford, 1990, pp. 112–19.

13 Population concentration: World Resources Institute, *World Resources 1990–91*, World Resources Institute, Washington DC, 1990; Linden, Olof, 'Human Impact on Tropical Coastal Zones', *Nature and Resources*, 26(4), pp. 3–11, 1990. Biological importance: World Resources Institute, *op. cit.*, pp. 132–4.

14 Meybeck *et al.*, *op. cit.*, p. 114; World Resources Institute, *op. cit.*, pp. 197, 182.

15 Wetlands extent and global loss: Maltby, Edward, *Waterlogged Wealth*, Earthscan, London, 1986, pp. 10, 90; Asian and African losses: MacKinnon, John and Kathy, *Review of Protected Areas System in the Afrotropical Realm*, and *Review of the Protected Areas System in the Indo-Malayan Realm*, both from International Union for the Conservation of Nature, Gland, 1986.

16 Asia and Africa: MacKinnon, *op. cit.*; Malaysia: *Keeping Options Alive*; Indonesia and Philippines: Linden, Olof, 'Human Impact on Tropical Coastal Zones', *Nature and Resources* 26(4), pp. 3–11, 1990; International Union for the Conservation of Nature Working Group on Mangrove Ecology, *Global Status of Mangrove Ecosystems*, IUCN, Gland, 1983, pp. 37–8; Niger and Indus: United Nations Environment Programme, *The State of the Marine Environment*, UNEP Regional Seas Reports and Studies no. 115, UNEP, Nairobi, 1990, p. 18.

17 World Resources Institute, *World Resources 1986*, World Resources Institute, Washington DC, 1986, pp. 150–1.

18 *Ibid.*, p. 151; Linden, Olof, 'Human Impact on Tropical Coastal Zones', *Nature and Resources* 26(4), pp. 3–11, 1990; United Nations Environment Programme, *Technical Annexes to The State of the Marine Environment*, UNEP Regional Seas Reports and Studies no. 114, UNEP, Nairobi, 1990, p. 537.

19 Dumping: United Nations Environment Programme, *The State of the Marine Environment*, UNEP Regional Seas Reports and Studies no. 115, UNEP, Nairobi, 1990, pp. 13–14; oil: Organization for Economic Cooperation and Development, *The State of the Environment*, OCED, Paris, 1991, pp. 73–5; radioactive waste: *ibid.*, p. 77.

20 Material on plastics from: United Nations Environment Programme, *Technical Annexes to The State of the Marine Environment*, UNEP Regional Seas Reports and Studies no. 114, UNEP, Nairobi, 1990, pp. 4–8; beach survey: Organization for Economic Cooperation and Development, *The State of the Environment*, OCED, Paris, 1991, p. 79.

21 Animal kills: World Resources Institute, *World Resources 1987*, World Resources Institute, Washington DC, 1987, p. 128; ghost-fishing:

United Nations Environment Programme, *Technical Annexes to The State of the Marine Environment*, UNEP Regional Seas Reports and Studies no. 114, UNEP, Nairobi, 1990, pp. 4–8.

22 The Great Stink: Mitchell, R. J., and Leys, M. D. R., *A History of London Life*, Penguin, London, 1963, p. 273.

23 United Nations Environment Programme, *op. cit.* note 21, pp. 389–90.

24 Helmer, R., quoted in Meybeck, Michael *et al.*, *Global Freshwater Quality*, United Nations Environment Programme, Blackwell Reference, Oxford, 1990, pp. 293–4.

25 Material on the Ganges is from: Centre for Science and Environment, *The State of India's Environment*, New Delhi, 1982, pp. 20–8, Varanasi: Centre for Science and Environment, *The State of India's Environment 1984–5*, New Delhi, 1986, p. 48.

26 United Nations Environment Programme, *The State of the Marine Environment*, UNEP Regional Seas Reports and Studies no. 115, UNEP, Nairobi, 1990, pp. 88–93.

15 Air pollution and climate change

1 From Benedict, Ruth, *Patterns of Culture*, Routledge & Kegan Paul, London, 1935, p. 144.

2 Global Environmental Monitoring System, *Forest Damage and Air Pollution*, Report of the 1988 Survey, United Nations Environment Programme, Nairobi, 1989, pp. 52, 63.

3 United Nations Environment Programme, *Environmental Data Report*, UNEP, Blackwell Reference, Oxford, 1989, pp. 48–50.

4 Sawyer, Jacqueline, *Acid Rain and Air Pollution*, World Wide Fund for Nature, Gland, Switzerland, 1989, pp. 16–18.

5 Rodhe and Herrera, *Acidification on Tropical Countries*, John Wiley, Chichester, 1988. Sawyer, *op. cit.*, p. 29.

6 Möller, D., 'Estimation of Global Man-Made Sulphur Emission', *Atmospheric Environment* 18(1), p. 24, 1984; World Resources Institute, *World Resources 1988–89*, World Resources Institute, Washington DC, 1988, p. 164.

7 Organization for Economic Cooperation and Development, *Environmental Indicators*, OECD, Paris, 1991, p. 21.

8 United Nations Environment Programme, *Environmental Data Report*, UNEP, Blackwell Reference, Oxford, 1989, pp. 14–15, 29, 30. Asia includes 25 per cent of USSR emissions, Europe the other 75 per cent.

9 OECD, *op. cit.*, p. 23.

10 World Resources Institute, *World Resources 1990–91*, World Resources Institute, Washington DC, 1990, p. 118.

11 Sawyer, *Acid Rain and Air Pollution*, World Wide Fund for Nature, Gland, Switzerland, 1989, p. 30.

12 Increase in production: UNEP, *Environmental Data Report 1989–90*, p. 28; atmospheric concentration: World Resources Institute, *World Resources 1990–91*, World Resources Institute, Washington DC, 1990, pp. 346–9.

13 World Resources Institute, *op. cit.*, p. 350.

14 United Nations Environment Programme, *Action on Ozone*, UNEP, Nairobi, 1990.

15 United Nations Environment Programme, *Environmental Effects Panel Report*, UNEP, Nairobi, 1989, pp. 39–48.

16 United Nations Environment Programme, *Economic Panel Report*, UNEP, Nairobi, 1989, p. 91.

17 The following section is based primarily on Houghton, J. T. *et al*, *Climate Change: The IPCC Scientific Assessment*, Cambridge University Press, Cambridge, 1990; concentrations, rates of increase, and life-spans p. xvi; relative warming impacts, pp. xxi, 8.

18 Beatty, J. Kelly and Chaikin, Andrew, *The New Solar System*, Cambridge University Press, 1990, p. 93.

19 Carbon dioxide data from Houghton, J. T. *et al.*, *Climate Change: The IPCC Scientific Assessment*, Cambridge University Press, Cambridge, 1990, ch. 1.

20 The figure for carbon dioxide emitted by deforestation depends on estimates for two main factors: the rate of deforestation; and the amount of carbon stored per hectare, which varies from one ecological zone to another. There is no common agreement on a figure for either of these factors. Carbon dioxide emissions are less exact than either of these, since they depend on multiplying the two together. Estimates for anthropogenic biological emissions of carbon dioxide vary by a factor of more than 4, between 600 million tonnes right up to 2600 million tonnes. The extent of regrowth of deforested areas, which also absorb carbon dioxide, is a further complicating factor.

21 Houghton, J. T., *op. cit.*, p. xvi.

22 *Ibid.*, p. xxi, 8.

23 *Ibid.*, p. xvi, xxi, 8.

24 *Ibid.*, pp. 19–23.

25 *Ibid.* Concentrations, rates of increase, and life-spans p. xvi; relative warming impacts, pp. xxi, 8.

26 Intergovernmental Panel on Climate Change, *IPCC First Assessment Report*, vol. I, August 1990.

27 Houghton, *op. cit.*, (note 19), pp. 12, 16, 301ff.

28 Stanley, Steven, *Earth and Life through Time*, W. H. Freeman, New York, 1986, p. 565.

29 Parry, Martin, *Climate Change and World Agriculture*, Earthscan, London, 1990, pp. 36–60.

30 Houghton, *op. cit.*, pp. 289–96.

31 Parry, *op. cit.*, p. 84.

32 Parry, *op. cit.*, p. 63; Intergovernmental Panel on Climate Change, *IPPC First Assessment Report*, vol. I, World Meteorological Association, Geneva, 1990, p. 14; Houghton, J. T. *et al.*, *Climate Change: The IPCC Scientific Assessment*, Cambridge University Press, Cambridge, 1990, p. xxiv.

33 Parry, *op. cit.*, pp. 63, 70.

34 Food and Agriculture Organization, *Production Yearbook 1989*, FAO, Rome, 1990, Table 3.

35 World Bank, *World Development Report 1990*, World Bank, Washington DC, 1990, pp. 182–3.

36 Production figures from Food and Agriculture Organization, *Production Yearbook 1990*; stock figures from FAO, *Food Outlook*, February 1991, both FAO, Rome.

37 Intergovernmental Panel on Climate Change, *IPPC First Assessment Report*, vol. I, World Meteorological Association, Geneva, 1990, p. 21.

38 Houghton, J. T. *et al.*, *Climate Change: The IPCC Scientific Assessment*, Cambridge University Press, Cambridge, 1990, pp. 296–300; Legget, Jeremy, ed,, *Global Warming: The Greenpeace Report*, Oxford University Press, Oxford, 1990, p. 147; Commonwealth Secretariat, *Climate Change: Meeting the Challenge*, Commonwealth Secretariat, London, 1989, p. 39.

39 Intergovernmental Panel on Climate Change, *IPPC First Assessment Report*, vol. I, World Meteorological Association, Geneva, 1990; Legget, *op. cit.*, pp. 116–48; Commonwealth Secretariat, *op. cit.*, p. 40.

40 Commonwealth Secretariat, *op. cit.*, pp. 55–6.

41 Sulphur dioxide emission levels taken from: Organization for Economic Cooperation and Development, *Environmental Indicators*, OECD, Paris, 1991, p. 21. (See appendix) p. 311; chlorofluorocarbon emissions from: UNEP, *Environmental Data Report 1989–90*, p. 28.

42 Organization for Economic Cooperation and Development, *Environmental Indicators*, OECD, Paris, 1991, p. 21.

43 Carbon dioxide emission levels from UN Environment Programme, *Environmental Data Report 1991/92*, Basil Blackwell, Oxford, 1991. See appendix.

44 Calculated from Food and Agriculture Organization, Economic and Social Policy Dept, *Country Tables 1990*, FAO, Rome, 1990. (See appendix.)

45 Livestock figures from *ibid*. See appendix.

46 Sawyer, Jacqueline, *Acid Rain and Air Pollution*, World Wide Fund for

Nature, Gland, 1989, p. 2; World Resources Institute, *World Resources 1988–89*, World Resources Institute, Washington DC, 1988, p. 165.

16   Bangladesh

1 Socioeconomic data are from Ahmad, Mohiuddin, *Feasibility Study on Hatia Embankment*, Oxfam, Dhaka, 1986.
2 Cited in Jansen, Eirik, *Rural Bangladesh: Competition for Scarce Resources*, University Press, Dhaka, 1987.
3 World Bank, *Bangladesh Fisheries Sector Review*, World Bank, Washington DC, 1990.
4 Navin, R. E., *The Agriculture Sector in Bangladesh: A Database*, USAID, Dhaka, 1989; Bangladesh Bureau of Statistics, *Statistical Pocket Book of Bangladesh*, Dhaka, 1990, pp. 121–5.
5 International Panel on Climate Change, *Climate Change: The IPCC Impacts Assessment*, Australian Government Publishing Service, Canberra, 1990, pp. 6–1 to 6–25.
6 Houghton, J. T. *et al.*, *Climate Change: The IPCC Scientific Assessment*, Cambridge University Press, Cambridge, 1990, pp. xi–xii.
7 Mahtab, F. U., *Effect of Climate Change and Sea-Level Rise on Bangladesh*, Report for Commonwealth Secretariat, 1989.
8 *Bangladesh Action Plan for Flood Control*, World Bank, Washington DC, 1989.
9 Mahtab, F. U., *Effect of Climate Change and Sea-Level Rise on Bangladesh*, Commonwealth Secretariat, London 1989; Houghton, J. T. *et al.*, *Climate Change: The IPCC Scientific Assessment*, Cambridge University Press, Cambridge, 1990, pp. xxv–xxvi.
10 I am indebted to Pramod Unia of Oxfam, who visited Hatia soon after the cyclone, for the data that follow.

17   Towards a general theory

1 Lattimore, Richmond, *Greek Lyrics*, Phoenix Press, Chicago, 1960.
2 See Bongaarts, John, 'A Framework for Analyzing the Proximate Determinants of Fertility', *Population and Development Review* 4(1), pp, 105–32, 1978.
3 Commoner, Barry, 'The Environmental Cost of Economic Growth', *Chemistry in Britain* 8(2), pp. 52–65, 1972.
4 Commoner, Barry, 'Rapid Population Growth and Environmental Stress', in *Consequences of Rapid Population Growth in Developing Countries*, Proceedings of United Nations Expert Group Meeting, August 1988, ESA/P/WP.110, United Nations, New York, 1989.

5   Pareto, Wilfredo, *Cours d'économie politique*, para 1047.

18   Sharing the blame

1   Income shares: *World Development Report 1991*, World Bank, Washington, 1991; fertilizers: FAO *Fertilizer Yearbook 1989*, FAO Rome, 1990; oil and gas: Hall, D. O. and Scurlock, D. M. O., 'The Contribution of Biomass to Global Energy Use', cited in United Nations Environment Programme, *Environmental Data Report*, UNEP, Blackwell Reference, Oxford, 1989, p. 426.

2   Industrial and hazardous wastes: Organization for Economic Cooperation and Development *The State of the Environment*, OECD, Paris, 1991, pp. 146; effluents: Meybeck, Michael *et al.*, *Global Freshwater Quality*, United Nations Environment Programme, Blackwell Reference, Oxford, 1990, p. 47; carbon dioxide emissions: Intergovernmental Panel on Climate Change, *IPCC First Assessment Report*, vol. I, August 1990, WGIII p. 8; chlorofluorocarbon emissions: World Resources Institute, *World Resources 1990–91*, World Resources Institute, Washington DC, 1990, pp. 346–9.

3   Population share from United Nations Population Division, *World Population Prospects 1990*, United Nations, New York, 1991.

4   Disparity calculated on the basis of average Northern share of 83·5 per cent, based on examples cited in text. Population increase from United Nations, *op. cit.*

5   Fertilizer changes from FAO, *Fertilizer Yearbook 1989*, FAO, Rome, 1990; carbon dioxide projection from Intergovernmental Panel on Climate Change, *IPCC First Assessment Report*, vol. I, August 1990, WGIII p. 8.

6   World Bank, *World Development Report 1991*, World Bank, Washington DC, 1991.

7   Water and waste examples from *Environment and Urbanization* 1(1), pp. 40–50, 63, 1989.

8   The approach is based on Durning, Alan, *Apartheid's Environmental Toll*, Worldwatch Paper 95, Worldwatch Institute, Washington DC, 1990, p. 25. Although *direct* energy use probably is not so skewed as income, when *indirect* use – via the additional products and services bought – is included, the assumption is probably a fair one. Data sources: 'Income Shares from World Bank', *World Development Report 1990*, World Bank, Washington DC, 1990, pp. 236–7; population: United Nations Population Division, *World Population Prospects 1990*, *op. cit.*; CO_2 emissions: World Resources Institute, *World Resources 1990–91*, World Resources Institute, Washington DC, 1990, pp. 346–7. CO_2 is given in terms of carbon equivalent. Only emissions from fossil

fuels have been included, since emissions from deforestation cannot be assigned on the basis of income groups.

9   Lesotho livestock ownership figures from *National Conservation Plan for Lesotho*, Ministry of Agriculture, Maseru, 1988, p. 14.

10 Rudel, Thomas K., *Population Growth and Environmental Degradation in Rural Areas of Developing Countries*, paper prepared for United Nations Population Division, New York, 1990.

11 World Bank, *World Development Report 1984*, World Bank, Washington DC, 1984, Box 6.1, p. 109.

12 United Nations Development Programme, *Human Development Report 1991*, Oxford University Press, Oxford, 1991, Table 10, p. 167.

13 On Africa see: World Bank reports: *Accelerated Development in sub-Saharan Africa*, World Bank, Washington DC, 1981, and *Sub-Saharan Africa: From Crisis to Sustainble Growth*, World Bank, Washington DC, 1989. On Europe: Organization for Economic Cooperation and Development, *Agricultural and Environmental Policies*, OECD, Paris, 1989.

14 United Nations Environment Programme, *Environmental Data Report 1989/90*, p. 425.

15 See Pearce, David *et al.*, *Blueprint for a Green Economy*, Earthscan, London, 1989.

16 Hardin, Garret, 'The Tragedy of the Commons', *Science* 162, pp. 1243–8, 1968.

17 Sandford, Stephen, *Management of Pastoral Development in the Third World*, Wiley, New York, 1983, pp. 118–27.

18 See Harrison, Paul, *The Third World Tomorrow*, Penguin Books, London 1983, pp. 11–15.

19 Jodha, N., 'Population Growth and Common Property Resources', in *Consequences of Rapid Population Growth in Developing Countries*, United Nations Expert Group meeting, 23–26 August 1988, ESA/P/WP.110, United Nations, New York, 1989, pp. 209–30.

20 Cubatão material based on World Resources Institute, *World Resources 1990–91*, World Resources Institute, Washington DC, 1990, p. 41; Satterthwaite, David and Hardoy, Jorge, *Squatter Citizen*, Earthscan, 1989, pp. 196–8.

19   Options for action

1   This figure is based on the following guesstimates of affluence, meaning possession of at least one or two expensive consumer durables other than a bike or radio:

Region	Total population	'Affluent' per cent	'Affluent' population
Low-income countries:	2950m	5 per cent	150 million
Lower-middle-:	682m	15 per cent	100 million
Upper-middle-:	423m	30 per cent	130 million
Developed:	1152m	90 per cent	1040 million

Population figures from: World Bank, *World Development Report 1991*, World Bank, Washington DC, 1991.

2 United Nations Population Division, *Long Range Population Projections*, United Nations, New York, 1991.

3 *Ibid.*

4 *Ibid.*

5 1980 projections from: United Nations Population Division, *Population Bulletin of the United Nations*, no. 14, 1982, p. 22.

6 Corley, Thomas, *Domestic Electrical Appliances*, Cape, London, 1966, pp. 16, 19.

7 Imran, Mudassar, and Barnes, Philip, *Energy Demand in Developing Countries*, World Bank Staff Commodity Working Paper no. 23, World Bank, Washington DC, 1990, pp. 15–16.

8 *Ibid.*, p. 17.

9 This and subsequent car data from: Motor Vehicle Manufacturers' Association, *World Motor Vehicle Data 1990*, Detroit, 1990; population growth from United Nations Population Division, *World Population Propects 1990*, United Nations, New York, 1991.

10 Poverty figure from: World Bank, *World Development Report 1990*, World Bank, Washington DC, 1990. Other aspects of deprivation from United Nations Development Programme, *Human Development Report 1991*, Oxford University Press, Oxford, 1991, Table 3.

11 See Harrison, Paul, *The Greening of Africa*, Penguin and Paladin, New York and London, 1987, pp. 255–77 for details of low-cost approaches to health and family planning.

12 A third path was proposed in the classic, Chenery, Hollis, *et al.*, *Redistribution with Growth*, Oxford University Press, Oxford, 1974. The idea was that the poor could be uplifted without dragging down the rich by a slightly skewed growth in which the incomes of the poorest rose fastest. This still involved faster total growth.

13 United Nations Development Programme, *Human Development Report 1991*, Oxford University Press, Oxford, 1991; Motor Vehicle Manufacturers' Association, *Facts and Figures '91*, MVMA, Detroit, 1991.

14 Motor Vehicle Manufacturers' Association, *Facts and Figures '91*, MVMA, Detroit, 1991.

15 On the potlatch see Benedict, Ruth, *Patterns of Culture*, Routledge & Kegan Paul, London, 1935, pp. 125–61.

16 Grübler, Arnulf and Nowotny, Helga, 'Towards the Fifth Kondratiev Upswing', *International Journal of Technology Management*, 5(4), pp. 431–71.

17 World Resources Institute, *World Resources 1990–91*, World Resources Institute, Washington DC, 1990, pp. 316–7.

18 *Ibid.*, p. 142–3.

19 *Ibid.*, Table 21.4.

20 Aulus Gellius, *Attic Nights*, ii.24.13–14; Dio Cassius, *Roman History*, lvii. 15.1.

21 Organization for Economic Cooperation and Development, *Environmental Indicators*, OECD, Paris, 1991, p. 21.

22 Food and Agriculture Organization, *Production Yearbook 1982* and *1990*, FAO, Rome, 1983 and 1990.

23 Goldemberg, José *et al.*, *Energy for a Sustainable World*, World Resources Institute, 1987, pp. 77–80.

24 Harrison, Paul, *The Greening of Africa*, Penguin and Paladin, New York and London, 1987, pp. 300–18.

25 Efficiency data from International Energy Agency, *Fuel Efficiency of Passenger Cars*, IEA, Paris 1991, pp. 12, 18, 25.

26 Motor Vehicle Manufacturers' Association, World Motor Vehicle Data 1990, Detroit, 1990.

27 Potential: World Energy Conference, *1989 Survey of World Energy Resources*, London, 1989; siltation: Mahmood, K. *Reservoir Sedimentation*, World Bank Technical Paper no. 71, World Bank, Washington DC, 1987, pp. 5–8.

28 Ogden, Joan and Williams, Robert, *Solar Hydrogen*, World Resources Institute, Washington DC, 1989, pp. 44–8; Organization for Economic Cooperation and Development, *Environmental Impacts of Renewable Energy*, The OECD Compass Project, OECD, Paris, 1988, p. 49; current cropland from Food and Agriculture Organization, *Production Yearbook 1989*, FAO, Rome, 1990.

29 OECD, *op. cit.*.

30 Ogden, Joan and Williams, Robert, *Solar Hydrogen*, World Resources Institute, Washington DC, 1989.

31 Bulatao, Rodolfo, *et al.*, *World Population Projections 1989–90*, World Bank, Washington DC, 1990, p. xx; United Nations Population Division, *Long-range World Population Projections*, United Nations, New York, 1991.

32 United Nations Population Division, *World Population Prospects 1990*, United Nations, New York, 1991.

33 United Nations Population Division, *Long-Range World Population Projections*, United Nations, New York, 1991.

34   Land requirements: in 1988 each person in the Far East (excluding China) had 0·18 hectares per person cropland. In China, where yields were probably as high as they could get without totally unforeseen technology breakthroughs, they had 0·09 hectares. The 0·15 figure represents a middle level, and compares with 0·546 hectares per person in developed countries in 1988. Land areas per person calculated from Food and Agriculture Organization, *Production Yearbook 1990*, FAO, Rome, 1991; and United Nations Population Division, *World Population Prospects 1990*, United Nations, New York, 1991. The figure for non-agricultural needs is from Higgins, Graham et al., *Potential Population Supporting Capacities of Lands in the Developing World*, FAO, Rome, 1982.

35   See Chapter 13 for sources on waste.

36   Houghton, J. T. *et al.*, *Climate Change: The IPCC Scientific Assessment*, Cambridge University Press, Cambridge, 1990, p. xviii.

37   Burkina Faso emissions: World Resources Institute, *World Resources 1990–91*, World Resources Institute, Washington DC, 1990, p. 346.

38   UK and USA current emissions: *ibid.*

39   Projections from: United Nations Population Division, *World Population Prospects 1990*, United Nations, New York, 1991.

40   Country fertility declines: *ibid.*

41   Paul Harrison, *The Third World Tomorrow*, Penguin, London, 1980, pp. 197–200.

42   Ministry of Health and Family Welfare, *Year Book 1988–89*, Government of India, New Delhi, 1990, pp. 102–3, 171–2.

43   United Nations Population Fund, *The State of World Population 1990*, New York, 1990. Urbanization and income: World Bank, *World Development Report 1991*, World Bank, Washington DC, 1991; agricultural workforce: Food and Agriculture Organization, *Production Yearbook 1989*, FAO, Rome, 1990.

44   World Bank, *World Development Report 1991*, World Bank, Washington DC, 1991; United Nations Development Programme, *Human Development Report 1991*, Oxford University Press, Oxford, 1991.

45   Infant mortality rates: World Bank, *World Development Report 1991*, World Bank, Washington DC, 1991.

46   United Nations Population Division, *Family Building by Fate or Design*, United Nations, New York, 1988.

47   United Nations Population Division, *Fertility Behaviour in the Context of Development*, United Nations, New York, 1987, pp. 224–5.

48   World Bank, *World Development Report 1991*, World Bank, Washington DC, 1991.

49   Boulier, Bryan, 'Family Planning Programmes and Contraceptive Availability', in Birdsall, Nancy, ed., *The Effects of Family Planning*

Programmes on Fertility, World Bank, Washington DC, 1985; Lapham, Robert and Mauldin, W. Parker, in Lapham, R., ed., *Organizing for Effective Family Planning*, Committee on Population, National Research Council, National Academy Press, 1987.

50   United Nations Population Fund, *The State of World Population 1990*, UNFPA, New York, 1990.

51   United Nations Population Division, *Levels and Trends of Contraceptive Use*, United Nations, New York, 1988, p. 26.

52   See United Nations Population Fund, *The State of World Population 1990*, New York 1990 for a summary of the UNFPA's review of effective family-planning programmes and supporting references.

53   United Nations Population Division, *Fertility Behaviour in the Context of Development*, United Nations, New York, 1987, p. 227.

54   Bulatao, Rodolfo, *World Population Projections 1989–90*, World Bank, Washington, DC, 1990.

55   For detailed analysis see *The State of World Population 1992*, United Nations Population Fund, New York, 1992. Three small outliers were excluded. r = − ·63, p < ·0001.

20   Towards the third revolution

1   This and following projections in this section are from: United Nations Population Division, *World Population Prospects 1990*, United Nations, New York, 1991 (up to 2025) and United Nations Population Division, *Long-Range World Population Projections: Two centuries of Population Growth*, United Nations, New York, 1991 (up to 2150).

2   *Ibid.*

3   Lugo, Ariel, 'Estimating Reductions in the Diversity of Tropical Forest Species', and Mooney, Harold, 'Lessons from Mediterranean Climate Regions', both in Wilson, E. O., ed., *Biodiversity*, National Academy Press, Washington DC, 1988.

4   Thomas, Keith, *Man and the Natural World*, Allen Lane, London 1983, pp. 254–68.

5   Wordsworth, *Prelude*, viii, 620–730; *Tintern Abbey*.

6   Thomas *op. cit.*, p. 266.

7   For a review of attitudes to nature in the world's religions see Regenstein, Lewis, *Replenish the Earth*, SCM Press, London, 1991. Regenstein considerably exaggerates the ecological concern of ancient religions. But he does show that every religion has some traditional material that can be emphasized to support the new attitudes to nature.

8 See Stone, Christopher, *Earth and other Ethics*, Harper & Row, New York, 1987.

9 See Chapter 13 on waste. The carbon balloon is based on an average life expectancy of 76 years and a carbon output of 3·12 tonnes per person per year in industrialized countries (Intergovernmental Panel on Climate Change, *IPCC First Assessment Report*, vol. I, August 1990, WGIII p. 8).

HAMLET SOURCES

	Page	Source		
One part wisdom	7	IV	iv	42
The o'ergrowth of some complexion	21	I	iv	27
What a piece of work is man	26	II	ii	80
Who would fardels bear	28	III	i	76
To grunt and sweat under a weary life	29	III	i	77
Bounded in a nutshell	38	II	ii	258
The fall of a sparrow	55	V	ii	221
The paragon of animals	73	II	ii	316
The grinding of the ax	88	V	ii	24
Abatements and delays	100	IV	vii	21
A sterile promontory	115	II	ii	307
A little patch of ground	126	IV	iv	18
To pay five ducats, five, I would not farm it	132	IV	iv	20
Quintessence of dust	140	II	ii	317
The interim's mine	156	V	ii	73
The quick of the ulcer	167	IV	vii	123
The drossy age	186	V	ii	191
A sea of troubles	195	III	i	59
A pestilent congregation of vapors	205	II	ii	31
One woe doth tread upon another's heel	210	IV	vii	164
Sorrows come not single spies	221	IV	v	78–9
The oppressor's wrong	221	III	i	71
Particular faults	255	I	iv	36
A couch for luxury	255	I	v	83
We defy augury	270	V	ii	220
This project should have a back or second	281	IV	vii	152–3
If it be not now, yet it will come	292	V	ii	224
I do not know why I yet live to say 'This thing's to do.'	292	IV	iv	43–6
That we would do, we should do when we would	297	IV	vii	117–8
The readiness is all	303	V	ii	224
The time is out of joint	305	I	v	188–9

INDEX

Note: key concepts are in bold capitals